Dynamics of marine sands

A manual for practical applications

Richard Soulsby

ThomasTelford

Published by Thomas Telford Publications, Thomas Telford Services Ltd,
I Heron Quay, London E14 4JD

URL: http://www.t-telford.co.uk

First published 1997

Distributors for Thomas Telford books are
USA: American Society of Civil Engineers, Publications Sales Department,
345 East 47th Street, New York, NY 10017-2398
Japan: Mauruzen Co. Ltd, Book Department, 3–10 Nihonbashi 2-chome,
Chuo-ku, Tokyo 103
Australia: DA Books and Journals, 648 Whitehorse Road,
Mitcham 3132, Victoria

Cover picture courtesy of HR Wallingford Ltd

A catalogue record for this book is available from the British Library

ISBN: 0 7277 2584 X

Typeset by Alden Group, Oxford
Printed and bound in Great Britain by Bookcraft (Bath), Limited

To Christine, Paul and my parents

Preface

The implications of sediment dynamics in the broadest sense are surprisingly far-reaching. Indeed, it is scarcely an exaggeration to say that the birth of civilisation in the fertile alluvial valleys of Mesopotamia, Egypt and China was dependent on sediment transport. In still earlier times, the deposition of sediments in valleys, lakes and seas laid down the geological sequence of sedimentary rocks which forms the basis of much of the sciences of geology, palaeontology and hydrocarbon exploration. The sedimentary processes that led to the fossilisation of plants and animals have also sometimes preserved wave or current generated ripple marks and other evidence of the hydraulic conditions in former epochs. Younger deposits show evidence of their depositional history in the form of pollen or foraminifera, and have preserved the artefacts that form the raw material of archaeology. From a geographical standpoint, the structure of the landforms and seas of the earth is largely determined by the erosion of rocks and their transport and deposition as sediments. Even on other planets the sediments may provide clues, as in the interpretation of river-like features as evidence for the former existence of water in the liquid state on Mars.

The scope of the present book is more limited than that sketched above, being restricted to non-cohesive sediments and mainly to the marine context, although some of the material can also be applied to estuaries and rivers. Its main focus is on those aspects of sediment dynamics that are of concern primarily to civil engineers, oceanographers, earth scientists and environmental scientists. The movement of sediment in rivers, estuaries and the sea is a subject of great practical importance for engineering applications, and it is also a fascinating and challenging area of academic research. However, the practical engineer looking for an urgent solution for a sediment-related project often finds that the results of the relevant academic research are published in unfamiliar language in journals and conference proceedings that are not readily to hand. Conversely, the academic researcher may have insufficient contact with the practical applications to decide which avenues of research are most in need of pursuit. My aim in this book has been to bridge the gap between the academic research and the practical applications, by summarising the research results in a unified form, by presenting worked examples and case studies, and by highlighting areas where understanding is deficient and error bounds are large. In doing so, I have sometimes made simplifications to which the purist might object, and in some cases I have taken liberties with re-casting research results into a common format, which the original authors might justifiably feel deviates to some degree from their own conventions. I hope the consequent gain in uniformity and usability outweighs any loss of precision.

Most of the sections of the book are divided into sub-Sections on *Knowledge* and *Procedure*. The research student can concentrate mainly on the *Knowledge*, perhaps supplemented by exercises based on the examples in *Procedure*, whereas the practising engineer may wish to just follow the *Procedure*, possibly backed up by reading associated parts of the *Knowledge*. Some deliberate repetition is included, and symbols are usually defined within each section, so as to make individual topics reasonably self-contained. Extensive cross-referencing is included where this is not possible.

I hope that in this way the book gives a convenient and usable introduction to sediment processes in a form that is accessible to a wide readership.

Richard Soulsby obtained degrees in physics and oceanography before specialising for 11 years in sediment transport research at the Institute of Oceanographic Sciences. Since 1985 he has worked on research and consultancy studies in this field at Hydraulics Research Wallingford, where he is currently head of the Marine Sediments Group.

Acknowledgements

This book grew out of an earlier report entitled 'Manual of Marine Sands'. In making this transformation I am indebted to a panel of invited reviewers of the earlier version, which included: Mr B. H. Rofe, Dr C. A. Fleming, Dr K. J. Riddell, Professor K. R. Dyer, Dr J. Nicholson, Dr Ping Dong, Mr C. W. Frith, Mr P. D. Hunter, Mr K. W. Bunker, Mr C. T. Erbrich, Mr P. D. Crews, Mr M. W. Owen, Mr R. Runcie, Mr D. Richardson and Mr P. B. Woodhead. In addition, valuable contributions were made by my colleagues at HR Wallingford: Dr R. J. S. Whitehouse, Mr N. P. Bunn and Mr J. S. Damgaard. I would like to convey my appreciation to all of the above scientists and engineers for their thorough and perceptive comments which led to considerable improvements in the book. I would also like to thank June Clarkson and Ruth Smith for their patient and careful typing of the manuscript.

Many of the topics of my own research quoted in the book were initially stimulated by a six-month period spent as a visiting fellow at the Norwegian University of Science and Technology (NTNU) in Trondheim. I am indebted to the Norwegian Council for Scientific and Industrial Research (NTNF) and to Professor Dag Myrhaug for making this possible.

The book is published on behalf of the Department of the Environment, Transport and the Regions who part funded its preparation. The views and information presented in the book are those of HR Wallingford and not necessarily those of the funding Department.

HR Wallingford is an independent specialist research, consultancy, software and training organisation that has been serving the water and civil engineering industries worldwide for over 50 years in more than 60 countries. We aim to provide appropriate solutions for engineers and managers working in:

- water resources
- irrigation
- groundwater
- urban drainage
- rivers
- tidal waters
- ports and harbours
- coastal waters
- offshore.

Address: Howbery Park, Wallingford, Oxon, OX10 8BA, UK
Internet: http://www.hrwallingford.co.uk

Notation

a, b, A, B	dimensionless coefficients in various formulae
$A = U_w T/(2\pi)$	orbital amplitude of wave motion at the bed
$b = \dfrac{w_s}{\kappa U_*}$	Rouse number (suspension parameter)
C	volume concentration of sediment (volume/volume)
C_{100}	drag coefficient applicable to current at 1 m
C_a	sediment reference concentration (volume/volume) at height z_a
C_D	drag coefficient applicable to depth-averaged current
C_M	mass concentration of sediment (mass/volume)
C_0	reference concentration (volume/volume) at the sea bed ($z = 0$)
cos	cosine
d	sieve diameter of grains
$D_* = \left[\dfrac{g(s-1)}{\nu^2}\right]^{1/3} d_{50}$	dimensionless grain size
d_n	grain diameter for which $n\%$ of the grains by mass is finer, e.g. d_{10}, d_{90}
d_{50}	median grain diameter
$d_{50,b}$	median diameter of grains in sea bed
$d_{50,s}$	median diameter of grains in suspension
d_{cr}	grain diameter which is just immobile for a given flow
e	2·718281828
$\exp(x) = e^x$	exponential function
f	Coriolis parameter
f_w	wave friction factor

f_{wr}	rough-bed wave friction factor
f_{ws}	smooth-bed wave friction factor
g	acceleration due to gravity $= 9 \cdot 81 \text{ ms}^{-2}$
h	water depth
H	height of water wave
H_{bk}	height of wave at breaking
H_0	deep-water wave height
H_{rms}	root-mean-square wave height
H_{s}	significant wave height
I	water surface slope (hydraulic gradient)
$k = 2\pi/L$	wave number of water waves
k_{s}	Nikuradse equivalent sand grain roughness
K_{I}	coefficient of permeability
K_{p}	specific permeability
L	wavelength of water wave
L_{p}	wavelength at the peak of the offshore wave spectrum
L_0	deep-water wavelength
$\ln$	natural logarithm (to base e)
$\log_{10}$	logarithm to base 10
p	pressure
p_{c}	probability distribution of current speed
p_{cw}	joint probability of current speed and wave orbital velocity
p_{H}	probability distribution of wave height
p_{w}	probability distribution of wave orbital velocity
q_{b}	volumetric bedload sediment transport rate transported per unit time per unit width
$q_{\text{B}} = q_{\text{b}}/(1 - \varepsilon)$	volume of settled-bed material (including pore-water) transported per unit time per unit width
$Q_{\text{b}} = \rho_{\text{s}} q_{\text{b}}$	mass sediment transport rate
q_{bx}	component of q_{b} travelling in the direction of the current
q_{by}	component of q_{b} travelling at right angles to the current, in the same sense as the angle ϕ
$q_{\text{b1/2}}$	wave half-cycle volumetric bedload sediment transport rate
Q_{LS}	longshore sediment transport rate integrated

	across surf-zone, in volume of sediment (excluding pore space) per unit time
q_s	suspended volumetric sediment transport rate
q_t	total volumetric sediment transport rate
q_{tc}	sediment transport rate due to a current alone
q_{tx}, q_{ty}	components of total volumetric sediment transport rate in positive x, y directions
$R_w = \dfrac{U_w A}{\nu}$	wave Reynolds number
$s = \rho_s/\rho$	ratio of densities of grain and water
sin	sine
sinh	hyperbolic sine
T	period of water wave
T_m	mean period of water waves
$T_n = (h/g)^{1/2}$	scaling period for waves
T_p	period at the peak of wave spectrum
$T_s = (\tau_{0s} - \tau_{cr})/\tau_{cr}$	transport parameter
T_z	zero-crossing period of water waves
t	time
tanh	hyperbolic tangent
U	horizontal component of water velocity
$\bar{U}$	depth-averaged current speed
$u_* = (\tau_0/\rho)^{1/2}$	friction velocity
u_{*f}	form-drag friction velocity
$u_{*m} = (\tau_m/\rho)^{1/2}$	mean friction velocity
$u_{*max} = (\tau_{max}/\rho)^{1/2}$	maximum friction velocity
u_{*s}	skin-friction velocity
u_{*t}	sediment-transport component of friction velocity
U_{10}	velocity at height of 0·1 m
U_{100}	current speed at a height of 1 m above the bed
$\bar{U}_a$ and $\bar{U}_b$	maximum and minimum values of the depth-averaged current speed through a tidal cycle
$\bar{U}_{cr}$	threshold depth-averaged current speed
U_{rms}	root-mean-square wave orbital velocity at sea bed
U_w	wave orbital velocity amplitude at sea bed
U_{wc}	bottom orbital velocity under wave crest

U_{wcr}	threshold wave orbital velocity
U_{wt}	bottom orbital velocity under wave trough
U_x, U_y	components of current in x, y directions
V_B	bulk velocity of water through bed
w_{mf}	minimum fluidisation velocity
w_s	settling velocity of isolated sediment grains
w_{sC}	settling velocity of grains in a dense suspension
x	horizontal coordinate
$X = \tau_c/(\tau_c + \tau_w)$	
	relative current strength
y	horizontal coordinate orthogonal to x
$Y = \tau_m/(\tau_c + \tau_w)$	
	dimensionless mean shear-stress
$Z = \tau_{max}/(\tau_c + \tau_w)$	
	dimensionless maximum shear-stress
z	height above sea bed
z_a	reference height near sea bed, at which reference concentration C_a is calculated
z_0	bed roughness length
α_b	angle between wave crests and shoreline at breaker line
β	angle of sloping bed to the horizontal
β_x	angle of sloping bed to the horizontal in direction of flow
β_y	angle of sloping bed to the horizontal in direction at right angles to the flow
δ	boundary-layer thickness
Δ_r	height of ripples
Δ_s	height of sandwaves
ε	porosity of bed
ζ	bed level, relative to an arbitrary datum
$\theta = \dfrac{\tau_0}{g(\rho_s - \rho)d}$	Shields parameter
θ_{cr}	threshold Shields parameter
θ_f	form-drag Shields parameter
θ_m	mean value of θ over a wave cycle
θ_{max}	maximum value of θ over a wave cycle
θ_s	skin-friction Shields parameter
θ_t	sediment-transport component of Shields parameter

θ_w	amplitude of oscillatory component of θ due to wave
κ	von Karman's constant $= 0.40$
λ_r	wavelength of ripples
λ_s	wavelength of sandwaves
μ	molecular viscosity
$\nu = \mu/\rho$	kinematic viscosity of water
π	3.141592654
ρ	density of water
ρ_B	bulk density of bed or suspension
ρ_s	density of sediment grains
σ	radian tidal frequency
$\sigma_g = (d_{84}/d_{16})^{1/2}$	geometric standard deviation of grain size
$\sigma_s = 0.5\,(d_{84}/d_{50} + d_{50}/d_{16})$	grain sorting parameter
τ	horizontal shear-stress in water column
τ_0	(total) bed shear-stress
τ_c	current-only bed shear-stress
τ_{cr}	threshold bed shear-stress for motion of sediment
τ_{0f}	form drag component of τ_0, due to bedforms
τ_m	mean bed shear-stress during a wave cycle under combined waves and currents
τ_{max}	maximum bed shear-stress during a wave cycle under combined waves and currents
τ_{0s}	skin-friction component of bed shear-stress
τ_{0t}	sediment-transport component of bed shear-stress
τ_w	amplitude of oscillatory bed shear-stress due to waves
$\tau_{\beta cr}$	threshold bed shear-stress on a bed sloping at angle β to the horizontal
ϕ	angle between current direction and direction of wave travel
ϕ	phi-size of grains
ϕ_i	angle of repose of sediment
ϕ_r	angle of final repose

$$\Phi_{x,y} = \frac{q_{bx,y}}{[g(s-1)d^3]^{1/2}}$$

dimensionless bedload transport rate, x and y components

ψ angle between current and upslope direction of a sloping bed

$$\Psi = \frac{U_w^2}{g(s-1)d}$$ wave mobility parameter

$\omega = 2\pi/T$ (absolute) radian frequency of waves

$\underset{\sim}{\cdot}$ horizontal vector quantity with components $(\cdot_x, \cdot_y)$

$\langle \cdot \rangle$ time average over many waves

$\langle \cdot \rangle_{LT}$ long-term mean (over months or years)

Contents

Illustrations

Introduction

1

1. Introduction

1.1. GENERAL

This book summarises, in an easily applicable form, the main processes determining the behaviour of sand in the sea. The results are intended to provide the tools with which practitioners can make calculations of sand behaviour in the context of engineering works. Although the main emphasis is on sediments in the sand range of grain size, many of the topics apply also to larger grains including gravels. Likewise, many of the results can be applied in estuaries and rivers, as well as in the sea.

It is a companion volume to the *Estuarine Muds Manual*, issued as HR Report SR 309 (Delo and Ockenden, 1992), and the *Manual of Sediment Transport in Rivers*, HR Report SR 359 (Fisher, 1993). The present manual updates, extends and replaces the earlier *Manual of Marine Sands*, HR Report SR 351 (Soulsby, 1994).

The aim of the book is to provide methods for calculating the various hydrodynamic and sediment dynamic quantities required for marine sediment transport applications in an easily accessible and unified form, with recommendations as to the most appropriate methods to use. It is intended primarily as a reference and 'how-to-do-it' book, and therefore does not include lengthy derivations and discussions. It will also be valuable for educational purposes, particularly when used in conjunction with the SandCalc software package for training exercises.

Several other excellent text books are available, including those by Sleath (1984), Dyer (1986), Fredsøe and Deigaard (1992), Nielsen (1992) and Van Rijn (1993). Many of the principles of marine sediment transport derive from methods used in rivers. Fluvial sediment transport is dealt with in the books by Graf (1984), Yalin (1977) and Raudkivi (1990), and the manual by Fisher (1993). In the wider context, coastal sediment transport is dealt with by Muir-Wood and Fleming (1981) and Horikawa (1988), and sandy features further offshore are dealt with by Stride (1982). Applications to beach management are dealt with in the manual by Simm *et al.* (1996). Many aspects of coastal management are dealt with in the US *Shore Protection Manual* (CERC, 1984). Coastal morphodynamic modelling is dealt with by de Vriend (in preparation), and scour around marine structures is dealt with by Whitehouse (1997).

As well as presenting existing formulae and methods for calculating the various quantities required, in many cases new formulae are presented as well. These derive from the author's recent research, and are believed to be both simpler and more accurate than previous formulae, on the basis of comparisons with large sets of measurements. The new methods include formulae for: the vertical structure of tidal currents, the drag coefficient for steady and tidal flows on a flat bed of sediment, wave orbital velocity at the sea bed, the friction factor for waves, bottom friction in combined wave and current flow, the settling velocity of sand grains, the threshold bed shear-stress of sand grains under currents and/or waves, the bedload and total transport rate of sand under currents and/or waves, and the longshore transport of sediment.

Most of the formulae are dimensionally homogeneous, so that any consistent set of units (e.g. SI units) can be used for the parameters they contain. In the few formulae which are not dimensionally homogeneous, the appropriate units are specified. Note that, in dimensionally homogeneous SI units, grain size must be given in metres (e.g. $200 \, \mu m = 2 \times 10^{-4} \, m$).

The book is divided into chapters covering fundamental properties of sand and water, hydrodynamics (calculating the properties of currents and/or waves which are relevant to sand transport), sand processes and transport (formulae for tackling most of the main classes of problem), and morphodynamics (how to use the sediment transport predictions to calculate patterns of erosion and accretion). A further chapter gives guidance on methods of handling the complex climate of waves and currents encountered in many practical problems. Each of the main chapters is structured into sections, each of which outlines in summary form first the state of knowledge on the topic and then the procedure for making an engineering calculation. In many cases the procedure is illustrated with worked numerical examples, and the book concludes with full worked case studies of a number of the most common classes of practical sediment problem.

1.2.   SANDCALC SOFTWARE PACKAGE

To facilitate the quick and accurate calculation of many of the quantities featured in the book, a software package SandCalc has been developed to complement the book. This gives easy access to over 70 of the equations and methods quoted in the book, in a Windows-based menu system for use on a PC.

The menu system is structured according to the same headings and sub-headings as are used in the book. Equations available in SandCalc are labelled 'SC' against the equation number in this manual, and the title of the equation as it appears in SandCalc is usually self-evident. Where necessary, some additional explanation of methods available through SandCalc are given in the appropriate place in the book.

Quantities calculated by SandCalc as outputs from one equation are automatically carried as inputs to other equations. Input and output values are checked to ensure they fall within an allowable range, and default values are given for some parameters.

Most of the worked examples and case studies given in the book can be calculated easily using SandCalc. In the worked examples, the results of each step of the calculations are rounded to three significant figures. The accumulation of errors leads in some cases to small differences between the numbers in the example and the value given by SandCalc. In these cases, the SandCalc value is the more accurate one.

1.3.   THE BEHAVIOUR OF MARINE SANDS

Sand transport plays a vital role in many aspects of coastal, estuarine and offshore engineering. The movement of sand influences: the construction of economically viable harbours (dredging costs for harbours and approach channels are often critical to viability), the construction of coastal power stations and refineries (sand may enter the cooling water intakes), coastal flood defence (integrity of beaches and offshore banks is crucial to dissipate wave attack), the loss or growth of amenity beaches (crucial to the success of many holiday resorts), the safety of offshore platforms and pipelines (sea bed scour can lead to toppling of platforms or breakage of pipelines), and many other applications.

Sand is conventionally defined as sediment having grain diameters in the range 0·062 to 2 mm. Finer sediments are classified as clays and silts (muds), and their properties are strongly influenced by electrochemical and biological cohesion. In mixed sediments, the effect of cohesion is important in determining the sediment properties if more than 10% of the sediment is finer than 0·062 mm. Such mixtures are more resistant to erosion than either a pure sand or a pure mud. Grains larger than 2 mm are classed as gravels. The permeability of gravels is an important factor in determining gravel behaviour, and a proportion of sand mixed with the gravel can reduce its permeability.

Sand in the sea may be moved by currents (tidal, wind or wave driven), or by waves, or very commonly by both currents and waves acting together. The sand is transported by the basic processes of entrainment, transportation and deposition (Figure 1a). These three processes take place at the same time and may interact with each other.

Entrainment takes place as a result of the friction exerted on the sea bed by the currents and/or waves, with turbulent diffusion possibly carrying grains up into suspension.

Transportation takes place by grains rolling, hopping and sliding along the bed in response to the friction, and, in the case of sloping beds, gravity. This is known as *bedload* transport, and is the dominant mode of transport for slow flows and/or large grains. If the flow is fast enough (or the waves large enough) and the grains fine enough, sand will be put into suspension up to a height of several metres above the bed, and carried by the currents. This mode of transport is known as *suspended load* and is often much greater than the bedload transport. In typical marine and estuarine situations, the predominant mode of transport is bedload for grains coarser than about 2 mm, and suspended load for grains finer than about 0·2 mm.

Deposition occurs when grains come to rest in bedload transport, or by settling out of suspension. Most of the time,

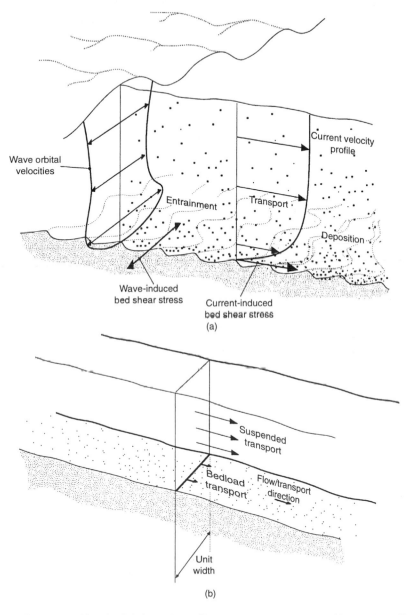

Figure 1. Sketch of (a) marine sediment transport processes (in practice all occur together) and of (b) plane normal to flow for defining sediment transport rate

entrainment of some grains upward into suspension and settling of other grains downward due to their weight take place simultaneously.

The *sediment transport rate* is defined as the 'amount' of sediment per unit time passing through a vertical plane of unit width perpendicular to the flow direction (Figure 1*b*).

The 'amount' of sediment may be measured by mass or by volume, so that in SI units the sediment transport rate is given in $\mathrm{kg\,m^{-1}\,s^{-1}}$ or $\mathrm{m^3\,m^{-1}\,s^{-1}}$ $(=\mathrm{m^2\,s^{-1}})$, respectively. More practical units such as tonnes per metre per day are also often used, with obvious conversions. Measurement in terms of 'immersed weight' is also sometimes used. The sediment transport rate in the sea has a magnitude and a direction, and hence is a vector quantity.

The rate of net *accretion* or *erosion* of an area of sea bed depends on the difference in the rates at which sand is entering and leaving the area. If sand is carried into a region faster than it is carried out, then the bed accretes; if the converse is true then it erodes. Even if the sediment transport rate is very large, the bed level will not change if the transport is equal over the whole area.

The operations required to make a prediction of the pattern of erosion and accretion in a study area are:

- Calculate the hydrodynamic distributions of currents and waves, and dependent quantities such as bed shear stress (bottom friction)
- Calculate the resulting distribution of sediment transport rates
- Calculate the distribution of erosion/accretion rates.

The emphasis in this book is on the physical process affecting sand, but, in addition, biological influences may play an important role, although to a lesser degree than they do for muds. Mucus secretions may help to bind grains together; conversely, worm casts may cause high spots which are more easily eroded by the currents; and re-working by burrowing organisms may mix the top 10 cm or so within 4 to 6 hours, thereby flattening current-formed ripple patterns. Little work has yet been done on quantifying biological influences on sediments; some of the most significant works are reviewed by Pender *et al.* (1994). Although there are at present few quantitative methods to deal with biological effects, the engineer should be aware that they may be present and can make predictions unreliable, especially in areas where the currents and waves are weak. In areas of strong currents or large waves, the main effects are usually hydrodynamic, and biological effects can justifiably be ignored.

1.4. INTRODUCTORY NOTE ON BED SHEAR-STRESSES

The effects of the hydrodynamic forcing agents (the currents and waves) on the sediment dynamics take place primarily through the friction they exert on the sea bed. This is expressed in terms of the *bed shear-stress*, which is the frictional force exerted by the flow per unit area of bed. Much of the book is devoted to methods of calculating bed shear-stresses, and to methods of determining their effects on the sediments. This introductory note serves to establish the relationships between the various types of bed shear-stresses, the various methods of expressing them, and the various contributions to them; and to define the notation that is used. Further details are given in the relevant sections of the main chapters, particularly Sections 3.4 and 4.6.

The *bed shear-stress*, τ_0, which has units of force per unit area $(\mathrm{N\,m^{-2}}$ in SI units), can also be written in units of velocity $(\mathrm{ms^{-1}}$ in SI units) as the *friction velocity* (or shear velocity), u_*, which is defined through the relationship:

$$\tau_0 = \rho u_*^2 \tag{1a}$$

$$u_* = (\tau_0/\rho)^{1/2} \tag{1b}$$

where ρ is the density of the water.

This conversion is made purely for mathematical convenience to save writing $(\tau_0/\rho)^{1/2}$ repeatedly, and the friction velocity does not correspond to a 'real' velocity in the flow (although it can be related to the turbulent fluctuations in the real velocity components). For many purposes, a dimensionless form of the bed shear-stress and its relationship to the sediment is used, namely the *Shields parameter*, θ, defined by

$$\theta = \frac{\tau_0}{g(\rho_s - \rho)d} \tag{2a}$$

or

$$\theta = \frac{u_*^2}{g(s - 1)d} \tag{2b}$$

where g = acceleration due to gravity
ρ = density of water
ρ_s = density of sediment grains
$s = \rho_s/\rho$ = ratio of densities of grain and water
d = diameter of sediment grains

The bed shear-stress generated depends not only the speed of the flow, but also on the roughness of the sea bed. This can be measured either by the *Nikuradse roughness*, k_s (which can be related to grain size), or by the *roughness length*, z_0 (which can be derived from the velocity profile). The two are connected through the relationship:

$$z_0 = k_s/30 \tag{3a}$$

$$k_s = 30z_0 \tag{3b}$$

The above notations apply to general, possibly time-varying, values of the bed shear-stress and the other quantities. However, it is also useful to distinguish between different generation mechanisms by using subscripts. The bed shear-stress may be generated by:

- a current, with the time-mean quantities denoted by τ_0, u_* and θ (i.e. the same as the general notation)
- a wave, with the amplitudes of the oscillatory quantities denoted by τ_w, u_{*w} and θ_w
- a combined wave and current motion, with the time-mean quantities over a wave cycle denoted by τ_m, u_{*m} and θ_m, and the maximum values of the quantities during a wave cycle denoted by τ_{max}, u_{*max} and θ_{max}.

Turbulent fluctuations cause large variations about the values given above, but for most practical purposes the above average values are sufficient. There may be a difference in the value of z_0 (and k_s) as seen by a current and by a wave, particularly if the orientations of current and wave to the crests of sand ripples on the bed are different, but, for simplicity, the same value of z_0 will be assumed to apply to both current and waves in this book.

The total bed shear-stress (τ_0) acting on the bed is made up from contributions due to:

- the *skin friction* (τ_{0s}) produced by (and acting on) the sediment grains

- the *form drag* (τ_{0f}) produced by the pressure field associated with the flow over ripples and/or larger features on the bed
- a *sediment-transport* contribution (τ_{0t}) caused by momentum transfer to mobilise the grains.

The three components can be added:

$$\tau_0 = \tau_{0s} + \tau_{0f} + \tau_{0t} \tag{4}$$

Similarly, three contributions to u_*, θ and z_0 can be identified, using the subscripts s, f and t to produce relationships like Equations (1), (2) and (3) for each contribution. Bed shear-stresses due to waves, or waves-plus-currents, can be subdivided into skin-friction, form-drag and sediment-transport contributions in the same way as that shown for currents. The complete set of notations used in this book for bed shear-stresses is given in Table 1.

An alternative commonly-used convention (e.g. Fredsøe and Deigaard (1992); Van Rijn (1993)) is to denote the components by primes, thus:

total $= \tau_0$, skin friction $= \tau_0'$,

form drag $= \tau_0''$, sediment transport $= \tau_0'''$.

Van Rijn (1993) refers to the skin-friction contribution as the "grain-related" contribution; this makes a useful distinction, as the usual method of calculating this contribution is by means of a logarithmic profile relationship or similar (see Section 3.1) with a value of z_0 (or k_s) related only to the grain size. The resulting value of bed shear-stress is not truly the skin friction which would be measured on the surface of a rippled bed, but it is of a similar size, and gives a practical conventional measure against which sediment responses can be correlated. Nielsen (1992) also emphasises this convention by using the notation $\tau_{2.5}$ for the skin-friction (grain-related) contribution obtained by setting $k_s = 2 \cdot 5 d_{50}$ (a conventional value) as the grain roughness used to compute this contribution.

If the bed is flat, and sediment transport is not intense, then Equation (4) simplifies to to $\tau_0 = \tau_{0s}$, and the distinction between total and skin-friction quantities is not necessary. This case is not uncommon for coarse grains (say $d_{50} > 0 \cdot 8$ mm), but finer sands are usually either rippled (possibly with larger bedforms) or have intense sheet-flow transport.

Table 1. *Notation for bed shear-stresses and related parameters*

Forcing agent	Contribution			
	Total	Skin friction	Form drag	Sediment transport
Current (or general) (mean values)	τ_0 u_* θ	τ_{0s} u_{*s} θ_s	τ_{0f} u_{*f} θ_f	τ_{0t} u_{*t} θ_t
Waves (amplitude of oscillatory values)	τ_w u_{*w} θ_w	τ_{ws} u_{*ws} θ_{ws}	τ_{wf} u_{*wf} θ_{wf}	τ_{wt} u_{*wt} θ_{wt}
Waves-plus-currents (mean values)	τ_m u_{*m} θ_m	τ_{ms} u_{*ms} θ_{ms}	τ_{mf} u_{*mf} θ_{mf}	τ_{mt} u_{*mt} θ_{mt}
Waves-plus-currents (maximum values)	τ_{max} u_{*max} θ_{max}	$\tau_{max,s}$ $u_{*max,s}$ $\theta_{max,s}$	$\tau_{max,f}$ $u_{*max,f}$ $\theta_{max,f}$	$\tau_{max,t}$ $u_{*max,t}$ $\theta_{max,t}$
All	z_0	z_{0s}	z_{0f}	z_{0t}

Note. The bed shear stress (τ_0), friction velocity (u_*) and Shields parameter (θ) are related through Equations (1)–(3). The contributions are linked through Equation (4).

The reason for including this note in some detail is that the various types of bed shear-stress often cause confusion. It is important to be aware that only the skin-friction contribution acts directly on the sediment grains, and it is therefore this contribution which is used to calculate the threshold of motion, bedload transport (with a few exceptions), and reference concentration or pick-up rate for grains in suspension. On the other hand, it is the total bed shear-stress that corresponds to the overall resistance of the flow (see Section 7.4), and determines the turbulence intensities which influence the diffusion of suspended sediment to higher levels in the water column (see Chapter 8). Failure to observe these distinctions would lead to serious errors in calculations.

1.5.   GENERAL PROCEDURE FOR SEDIMENT STUDIES

The following sets out a general procedure to follow for solving a wide variety of sediment-related problems. There is, in fact, no standard or widely-accepted procedure established for such problems, and the following is purely a personal approach suggested by the author. The procedure should be adapted to the specific problem, and rarely (if ever) will every step need to be taken. In some steps, rather simplified rules are given, and specific Sections should be referred to if a more detailed account is required. Very often the input information will be incomplete, and assumptions must be made or default values taken. However, default values should be used only if no site-specific values are available. The default values stated are for typical sea conditions around the UK. Most of the steps shown can be calculated using the software package SandCalc.

1. Gather as much background information as possible about the *study area*. Examine a large-scale chart (and/or map, if applicable) of the area, looking for information about:

   - bathymetry and water depths; sediment transport is usually greatest in shallow water
   - sediment type and mobility, e.g. from spot markings of Quality of the Bottom (Section 2.2); evidence of spits (indicate direction of net longshore transport), tombolos, channels and sandbanks, rocky outcrops, muddy bays
   - current speed and direction, e.g. from 'tidal diamonds'; current speeds in excess of about $0.4\,\mathrm{m\,s^{-1}}$ will have a significant effect on sediment transport
   - wave climate; exposure to locally-generated waves (look at distance to nearest coastline in all directions from study site), and swell waves (look for direct pathways to the open ocean).

In British waters, the British Geological Survey publishes charts of Seabed Sediments (BGS, 1987) showing detailed distributions of sediments types, and evidence of sediment mobility and direction of transport in terms of sandwaves (asymmetry indicates transport direction), sandbanks, iceberg ploughmarks (imply negligible sediment transport over millennia), and others.

If possible, make a site visit. Take sediment samples, noting degree of nonuniformity both across the site and vertically (e.g. a thin veneer of sand over a gravel or clay substrate). Build-up of sediment at groynes and other structures indicates transport direction. Growth of weed or barnacles on pebbles indicates low mobility. Observe current patterns (oranges make good surface drifters) and wave patterns. Talk to local people, especially fishermen.

2. Obtain basic information about the *water properties*:

- Depth, h. In tidal areas, if the tidal range is much less than the mean depth (i.e. the difference between the elevations of Mean Sea Level and the sea bed), use mean depth throughout. Otherwise, match depth and current variations through the tidal cycle.
- Temperature (default $= 10\,°C$) and salinity (default $=$ 35 ppt). Use annual averages of near-bottom values, unless seasonal variations are important.
- Calculate water density, ρ (default $= 1027\,\mathrm{kg\,m^{-3}}$) and kinematic viscosity, ν (default $= 1.36 \times 10^{-6}\,\mathrm{m^2\,s^{-1}}$) (Section 2.1).
- Take acceleration due to gravity as $g = 9.81\,\mathrm{m\,s^{-2}}$.

3. Obtain basic information about the *bed material* (Section 2.2):

- Take samples of the top 5 cm or so of the bed using a grab, borehole, corer or by hand. Preferably take six to eight samples over the study area to assess variability; these can either be averaged, or one typical sample chosen. Use BGS or Admiralty charts if no sample is available.
- Obtain the grain-size distribution of the sand and gravel fraction of the bed samples by sieving. The median grain diameter d_{50}, must be derived, and preferably as many as possible of d_{10}, d_{16}, d_{35}, d_{65}, d_{84} and d_{90}, depending on what methods are to be used subsequently. If some of these values are not known, they can be estimated by assuming a log-normal grain-size distribution between known points.
- If d_{10} is less than 0.06 mm, then the sediment may exhibit cohesive properties and the methods given for sandy sediments will be unreliable. Test a moist sample to see if it falls apart (non-cohesive) or clings together (cohesive). The presence of ripples on the bed generally indicates a relatively noncohesive sediment. If the sediment is slightly cohesive, use the sand methods but treat the results with

caution. If it is very cohesive, then it must be treated as a mud, using methods given by, for example, Delo and Ockenden (1992).

- If most of the sediment is quartz or similar minerals, the density can be taken as a default value of $\rho_s = 2650 \, \text{kg m}^{-3}$. Otherwise, measure the density of the sediment (e.g. use a specific gravity bottle). Calculate density ratio $s = \rho_s/\rho$ (default $= 2.58$). If the density is very different from $2650 \, \text{kg m}^{-3}$, or grains are very angular or contain holes (e.g. shell fragments, coral sand, volcanic sand), then hydraulic measurements (e.g. threshold of motion; settling velocity) should be made. If less than 30% of the material is shell fragments, make calculations just for the quartz fraction and assume the shell is hydraulically similar.
- If sloping bed effects are important in the problem, measure the angle of repose of sediment ϕ_i (default $= 32°$). This can be done by measuring the sideslope angle of a conical pile of sediment formed by pouring a sediment sample underwater from a small height.
- Calculate for $d = d_{50}$:
  dimensionless grain size D_* (Equation (75))
  threshold Shields parameter θ_{cr} (Equation (77))
  threshold shear-stress τ_{cr} (Equation (74))
  settling velocity w_s (Equation (102)).

4. Decide on *prevailing flow conditions*, by consulting a chart and/or visiting the site:

- Current alone (e.g. river; estuary with bar at entrance and protected from local winds; offshore sites deeper than about 40 m). Go to Step 5.
- Waves alone (e.g. a coastal or lake-side near-shore zone in a depth of less than 5 m). Go to Step 6.
- Combined waves and currents (e.g. coastal and offshore sites in depths between 5 and 40 m; estuaries exposed to the open sea or with long and/or wide stretches exposed to strong winds). Go to Step 7.

5. *Current alone*

- Decide whether to use a single design current speed or a probabilistic distribution of speeds (Chapter 12). For long-term averaged calculations a probabilistic approach is recommended.

- Obtain measures of current speed through one tidal cycle each of mean spring and mean neap tides. Use a current meter, or a numerical model, or Admiralty Chart tidal diamonds, or Tidal Streams Atlas. For most of the subsequent formula it is the *depth-averaged* current $\bar{U}$ which is required. This is the quantity which would be predicted by a 2DH numerical model, and is approximately the value measured by a current meter mounted at a height of $0.32 \times$ depth. For current meters at other heights, and for Admiralty Tidal Streams (which give water surface values), convert to $\bar{U}$ using Equation (28).
- Calculate skin-friction bed shear-stress, τ_{0s}, friction velocity, u_{*s}, and Shields parameter, θ_s, (Section 3.3), using $d = d_{50}$.
  - if $\theta_s < \theta_{cr}$, then bed is immobile (assume rippled)
  - if $\theta_{cr} \leq \theta_s \leq 0.8$, then bed is mobile and rippled and/or duned
  - if $\theta_s > 0.8$, then bed is mobile and flat with sheet-flow
  - if $u_{*s} \leq w_s$, then no suspension
  - if $u_{*s} > w_s$, then sediment is suspended.

Similar calculations for other grain-size percentiles (d_{10}, d_{16}, d_{35}, etc.) will determine which grain-size fractions will move as bedload or in suspension, or be immobile. If $d_{50} > 0.8\,\text{mm}$, then ripples do not form.

Perform some or all of the following calculations as appropriate:

- Calculate height and wavelength of ripples and dunes, if present (Section 7.2).
- Calculate effective total roughness, z_0, and total bed shear-stress, τ_0, friction velocity, u_*, and Shields parameter, θ, either from dimensions of bed features or via an alluvial friction method (Section 7.4).
- Calculate median suspended grain size $d_{50,s}$ (Section 8.2), and corresponding settling velocity, w_s (Section 8.3). Use these to calculate suspended sediment concentration at desired heights (Section 8.4).
- Calculate bedload transport rate (Section 9.2), taking account of bed slope if appropriate.

- Calculate total sediment transport rate (Section 10.2).
- If the sediment transport rate has been calculated at a spatial array of points, calculate the pattern of erosion and deposition (Section 11.1).

6. *Waves alone*

- Decide whether to use a single design wave or a probabilistic distribution of wave heights, periods and directions (Chapter 12). For long-term averaged calculations a probabilistic approach is recommended.
- Obtain an $H_s - T_z$ scatter plot or similar, with directional information if appropriate (Section 4.2). This could be from *in situ* measurements with a wave-buoy, pressure-recorder, ship-borne wave recorder, wave-staff or visual observations; historical measurements nearby using these methods, or ship's-log observations; or by hindcasting computationally from wind records. If only H_s is available, estimate T_z using Equation (49). Calculate other measures (H_{rms}, T_p, equivalent monochromatic H and T) using the relationships in Section 4.2, as required.
- Check that wave height does not exceed height for wave breaking for that period and water depth (Section 4.7).
- Calculate bottom orbital velocity amplitude, U_w, if an equivalent monochromatic wave is taken, and/or root-mean-square bottom orbital velocity, U_{rms}, if a spectral approach is used (Section 4.4). The choice depends on the requirements of the subsequent formulae.
- Calculate amplitude of skin-friction bed shear-stress, τ_{ws}, friction velocity, u_{*ws}, and Shields parameter, θ_{ws}, using $d = d_{50}$ (Section 4.5):
  - if $\theta_{ws} < \theta_{cr}$, then bed is immobile (assume rippled)
  - if $\theta_{cr} \leq \theta_{ws} \leq 0.8$, then bed is mobile and rippled
  - if $\theta_{ws} > 0.8$, then bed is mobile and flat with sheet flow
  - if $u_{*ws} \leq w_s$, then no suspension
  - if $u_{*ws} > w_s$, then sediment is suspended.

Similar calculations for other grain-size percentiles (d_{10}, d_{16}, d_{35}, etc.) will determine which grain-size fractions will move as bedload or in suspension or be immobile.

Perform some or all of the following calculations as appropriate:

- Calculate height and wavelength of ripples, if present (Section 7.3).
- Calculate effective total roughness, z_0, from ripple dimensions. Calculate amplitudes of total bed shear-stress, τ_w, friction velocity, u_{*w}, and Shields parameter, θ_w (Section 4.6).
- Calculate median suspended grain size (Section 8.2), and corresponding settling velocity, w_s (Section 8.3). Use these to calculate suspended sediment concentration at desired heights (Section 8.5).
- Calculate wave-induced bedload transport rate (Section 9.3).
- Calculate total wave-induced sediment transport rate (Section 10.3).
- Calculate long-shore sediment transport rate (Section 10.5).

7. Combined waves and currents

- Decide whether to use a design-current + design-wave approach, or joint probabilistic distribution of currents and waves (Chapter 12). For long-term averaged calculations a probabilistic approach is recommended.
- Obtain measures of current speed, as in Step 5.
- Obtain wave heights, periods and directions, and convert to bottom orbital velocities, as in Step 6.
- Calculate mean (τ_{ms}) and maximum ($\tau_{max,s}$) values of the skin-friction bed shear-stress over a wave cycle (Section 5.3). Convert to mean (u_{*ms}) and maximum ($u_{*max,s}$) friction velocities, and mean (θ_{ms}) and maximum ($\theta_{max,s}$) Shields parameters:
  - if $\theta_{max,s} < \theta_{cr}$, then bed is immobile (assume rippled)
  - if $\theta_{cr} \leq \theta_{max,s} \leq 0 \cdot 8$, then bed is mobile and rippled
  - if $\theta_{max,s} > 0 \cdot 8$, then bed is mobile and flat with sheet flow
  - if $u_{*max,s} \leq w_s$, then no suspension
  - if $u_{*max,s} > w_s$, then sediment is suspended.

Similar calculations for other grain-size percentiles (d_{10}, d_{16},

d_{35}, etc.) will determine which grain-size fractions will move as bedload or in suspension or be immobile.

Perform some or all of the following calculations as appropriate.

- Calculate height and wavelength of current ripples (Section 7.2) and wave ripples (Section 7.3). Select the one with the larger height.
- Calculate effective total roughness, z_0, from ripple dimensions.
- Calculate mean (τ_m) and maximum (τ_{max}) values of the total bed shear-stress over a wave cycle (Section 5.3). Convert to mean (u_{*m}) and maximum (u_{*max}) friction velocities, and mean (θ_m) and maximum (θ_{max}) Shields parameters.
- Calculate median suspended grain size (Section 8.2), and corresponding settling velocity, w_s (Section 8.3). Use these to calculate suspended sediment concentration at desired heights (Section 8.6).
- Calculate mean bedload transport rate (Section 9.4).
- Calculate mean total sediment transport rate (Section 10.4).
- If the mean sediment transport rate has been calculated at a spatial array of points, calculate the pattern of erosion and deposition (Section 11.1).

The methods described in the above procedure, and indeed in most of the book, relate to calculations at a local point – say, the water column above one square metre or so of sea bed. Practical studies often require predictions over a more extended study area. For some classes of problem the procedure can be applied at every point of interest in the study area, as described in Chapters 11 and 12.

More generally, however, methods of predicting conditions in an extended area fall into the following three categories:

- observational experience, gathered in the study area and in similar areas, in which contemporary and historical evidence, supported by broad-based measurements, is interpreted by an experienced practioner (e.g. a geomorphologist) to provide an understanding of the behaviour of the study area and its likely response to engineering works
- physical models, which are small-scale models of the study area built in a laboratory to simulate either the hydrodynamic

current and/or wave distributions over a solid moulded bed, or the sediment dynamics and morphodynamics using appropriately scaled natural or lightweight sediments

- numerical models, which are computer-based solutions of the governing equations for the hydrodynamics, sediment dynamics and morphodynamics, that simulate the distributions of these quantities over a finite-difference or finite-element grid of the study area.

Each method is a specialised field of expertise in its own right, and all the methods require a background knowledge of the types of principles presented in this book for their proper implementation. A brief review, with some indications of the most appropriate choice of method, was given by Soulsby (1993), and a much more detailed reference book covering all the methods was compiled by Abbott and Price (1994).

1.6. ERRORS AND SENSITIVITIES

Sediment transport is still an inexact science. This is because it depends on many complicated and interactive processes, and a number of the processes are often not measured or are imperfectly understood. Biological effects, sediments with a wide range of grain size components, and the time history effects of ripple and sandwave dimensions being dependent on previous events, come into the latter category. In addition, it is speculated that a strongly non-linear process such as sediment morphodynamics may exhibit chaotic behaviour (in the mathematical sense) in the same way that the weather does. The engineer should therefore appreciate that even the best available predictions for sand transport will have a wider margin of error than would be expected in other branches of engineering and science. For example, in river engineering, the best available methods can only predict sediment transport rates to better than a factor of 2 in about 70% of cases. In the much more complicated environment of the sea, the best methods may not be able to achieve much better than an accuracy of a factor of 5 in sediment transport rates for 70% of cases. This can be improved considerably by making use of site-specific data, such as the infill rate of a dredged trial trench, to calibrate a formula for site-specific use. Despite this level of inaccuracy, as with weather forecasting, clients would usually prefer an imperfect answer to none at all.

Error and uncertainty arise for one or more of the following reasons: a prediction method not including all of the relevant processes involved; incomplete understanding or imprecise formulation of the processes which have been included; use of prediction methods outside their range of validity (particularly empirical methods); errors in measuring the input values; errors in measuring the 'observed' output values.

When making predictions it is therefore prudent to perform a sensitivity analysis to estimate:

• errors in outputs resulting from uncertainties in inputs,
• differences between prediction methods.

Typical uncertainties in some of the important input parameters are:

• density of water, ρ, $\pm 0.2\%$
• kinematic viscosity of water, ν, $\pm 10\%$
• sediment density, ρ_s, $\pm 2\%$
• grain diameters, d_{10}, d_{50}, d_{90}, etc., $\pm 20\%$
• water depth, h, $\pm 5\%$
• current speed, $\bar{U}$, $\pm 10\%$
• current direction, $\pm 10°$
• significant wave height, H_s, $\pm 10\%$
• wave period, T_z, $\pm 10\%$
• wave direction, $\pm 15°$.

These uncertainties represent a combination of instrumentation errors, and representativeness of a single value over a large site and long time. The uncertainties in water density and viscosity arise from seasonal or spatial variations in water temperature and salinity. Uncertainties in sediment properties arise from spatial and temporal variations in bed composition. Uncertainties in water depth and current arise from tide to tide variability, in addition to current meter errors. Uncertainties in wave properties arise from measurement and analysis techniques, and inter-annual variations in wave climate.

Some examples of the uncertainty of outputs that result from uncertainties in the inputs are:

• the concentration of suspended sediment at a height of 1·2 m predicted by Equation (115) for current speed $= 0\cdot44\,\mathrm{ms}^{-1}$, wave orbital velocity $= 0\cdot83\,\mathrm{m\,s}^{-1}$ and depth $= 13\,\mathrm{m}$ varies from $3\cdot0\,\mathrm{kg\,m}^{-3}$ to $0\cdot46\,\mathrm{kg\,m}^{-3}$ if the median grain size of suspended sediment is varied from $0\cdot10\,\mathrm{mm}$ to $0\cdot12\,\mathrm{mm}$.

- the sediment transport rate predicted by Equation (136) for sediment of grain size $d_{50} = 0\cdot2\,\mathrm{mm}$ in $6\,\mathrm{m}$ water depth with a $0\cdot75\,\mathrm{m\,s}^{-1}$ current varies from $0\cdot52 \times 10^{-3}\,\mathrm{m}^2\,\mathrm{s}^{-1}$ to $1\cdot0 \times 10^{-3}\,\mathrm{m}^2\,\mathrm{s}^{-1}$ if the significant wave height is varied from $2\cdot0$ to $2\cdot5\,\mathrm{m}$.

Differences between different prediction methods are relatively small for the well-understood parameters, but much larger for the less well-understood parameters. For example, the differences between methods are generally less than:

- 1% for water density and viscosity
- 10% for wave orbital velocity, and grain settling velocity
- 20% for threshold bed shear-stress, threshold current speed, threshold wave height
- 50% for mean and maximum bed shear-stresses under combined waves and currents
- a factor of 3 for suspended sediment concentrations and sediment transport rates under currents
- a factor of 5 for suspended sediment concentrations and sediment transport rates under waves plus currents

Uncertainties also arise in the procedures to be followed in applying a given method: e.g. how to handle a cohesive fraction of the sediment, if present; how to handle a shell fraction; what representative wave height and period to use to represent a spectrum of waves; how to represent the wave and current climate over long periods. Guidelines are given where possible in this book, although there are no international standards or even generally accepted approaches to follow. Different decisions on such procedures by different practitioners could easily lead to 50% differences in results, even if they all use the same prediction method.

Properties of water and sand

2. Properties of water and sand

2.1. DENSITY AND VISCOSITY OF WATER

Knowledge

The density of sea water generally decreases with temperature and increases with salinity. Fresh water has a maximum density at about 4 °C. The maximum density of sea water occurs at about −1·9 °C. Density is tabulated against temperature and salinity by Chen *et al.* (1973) and Myers *et al.* (1969). These tabulations have been used to prepare curves of density versus temperature and salinity (Figure 2).

A suspension of sediment also increases the effective bulk density of the water. The density effect of suspended sediment can dampen turbulence, or produce a turbidity current on a sloping bed.

The kinematic viscosity ν is a molecular property of water that is defined for a laminar flow by the relationship

$$\tau = \rho\nu\frac{dU}{dz} \tag{5}$$

where τ = horizontal shear-stress at height z
ρ = water density
U = horizontal velocity at height z
z = vertical coordinate

The molecular viscosity μ ($= \rho\nu$) is also commonly encountered, but ν is more widely used for sediment transport purposes. The kinematic viscosity of water decreases with temperature and increases with salinity. Kinematic viscosity is tabulated against temperature and salinity by Chen *et al.* (1973) and Myers *et al.* (1969). These tabulations have been used to prepare curves of viscosity versus temperature and salinity (Figure 3).

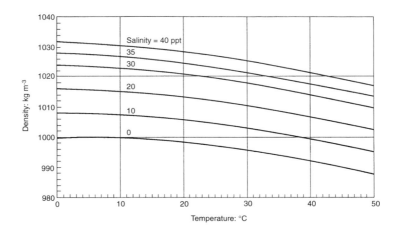

Figure 2. Density of water

A suspension of sediment increases the effective viscosity μ_e for *bulk* flow of water (e.g. through pipes), according to the relationship

$$\mu_e = \frac{\mu}{1 - 2 \cdot 5C} \tag{6}$$

where C is the concentration by volume. However, for flow around individual grains (e.g. to calculate settling velocity), the unmodified viscosity should be used.

Procedure

To calculate the density and kinematic viscosity of water:

1. Measure or predict the temperature and salinity of the water. For fresh water, salinity $= 0$. For typical sea water, salinity $= 35$ ppt.

2. *Example 2.1. Density and viscosity of water*

Temperature in degrees C.	10
Salinity in parts per thousand (ppt).	35

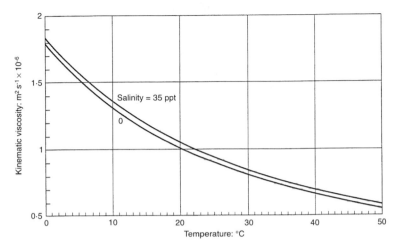

Figure 3.   Kinematic viscosity of water

Interpolate for density using the
curves in Figure 2 to obtain water
density.

$1027 \ \mathrm{kg\,m^{-3}}$

Interpolate for viscosity using
the curves in Figure 3 to obtain
kinematic viscosity.

$1.36 \times 10^{-6} \ \mathrm{m^2\,s^{-1}}$

3. For many purposes, the density of sea water can be taken
   as a default value of $1027 \ \mathrm{kg\,m^{-3}}$ (temperature = 10 °C;
   salinity = 35 ppt). This corresponds approximately to the
   annual mean bottom temperature and salinity of the seas
   around the UK (MAFF, 1981).

4. To calculate the density with suspended sediment, refer to Section
   2.3. To calculate the bulk effective viscosity with suspended
   sediment, use Equation (6).

2.2.   BED MATERIAL

Knowledge

Sediment grains are classified according to their diameter into
clays, silts, sands, granules, pebbles, cobbles and boulders. Clays

and silts are collectively called mud, and granules, pebbles and cobbles are collectively called gravel. Water-worn, rounded gravel is often referred to as shingle. The most commonly used classification is the Wentworth scale (Figure 4). Sand grains are sometimes measured in microns (µm) and sometimes in millimetres. Geologists use the phi-scale defined by

$$\phi = -\log_2 d \tag{7a}$$

$$d = 2^{-\phi} \tag{7b}$$

where d is the grain diameter *in millimetres*, and $\log_2$ is the logarithm to base 2. Conversion from ϕ to d is given in Figure 4.

Natural sands always contain a mixture of grain sizes, some of which may go outside of the sand range. The most common method of measuring the grain-size distribution is by sieving using a stack of sieves whose meshes decrease in size down the stack by a set ratio, often equivalent to a half-phi or a quarter-phi. The standard technique for washing, drying and sieving sands is described in BSI specifications (1967, 1986). In this book we will use the notation 'd' to correspond to the grain size obtained by sieving.

The grain-size distribution is usually presented as a cumulative curve showing the percentage by mass of grains smaller than d, versus d. An example is shown in Figure 5. The sediment is often characterised by its median sieve diameter d_{50} (the diameter for which 50% of the grains by mass is finer). More generally, the notation d_n indicates the grain diameter for which $n\%$ of the grains by mass is finer. The most commonly used percentiles are (in order of increasing size): d_{10}, d_{16}, d_{35}, d_{50}, d_{65}, d_{84}, d_{90}.

If grain diameter is denoted simply by 'd' in a prediction formula, it can usually be taken to be d_{50} of the bed material. However, in calculations of the vertical distribution of suspended sediment concentration, it is better to base the settling velocity of the grains on d_{50} of the material in suspension, which is often rather smaller than d_{50} of the bed material.

The spread of sizes is indicated by the 10 and 90 percentile sizes (d_{10} and d_{90}) or, alternatively, d_{16} and d_{84}. A commonly used measure of the degree of sorting of a sediment is the *geometric standard deviation* σ_g:

$$\sigma_g = \sqrt{d_{84}/d_{16}} \tag{8}$$

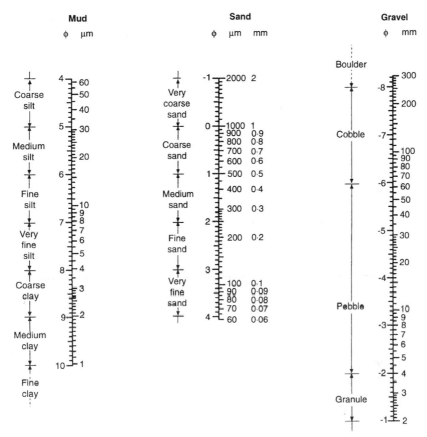

Figure 4. Conversion chart from phi units to microns (μm) and mm (Wentworth grain size scale)

Sediment samples are classed as well sorted if they contain a narrow range of grain sizes and well mixed if they contain a wide range. Dyer (1986) gives a detailed discussion of various methods of describing the nature of a sediment mixture. As a rough guide, if a sediment sample has $d_{84}/d_{16} < 2$ (or $d_{90}/d_{10} < 2.4$), then it is well sorted, whereas if it has $d_{84}/d_{16} > 16$ (or $d_{90} > 35$), then it is well mixed. The example in Figure 5 has $d_{10} = 0.23$ mm and $d_{90} = 2.6$ mm, so that $d_{90}/d_{10} = 11$, and it thus has an intermediate sorting.

Many mixed sediments have a grain-size distribution that approximates to a *log-normal distribution*; that is, the logarithm of the grain size has an approximately normal (Gaussian) frequency distribution by weight. The cumulative grain-size distribution (% finer by weight) of a log-normal distribution plots as a straight line on probability graph paper if the abscissa is in phi units. In phi units, the mean of the normal distribution is $-\log_2(d_{50})$ and the standard deviation is $\log_2(\sigma_g)$. If only d_{50} and a measure of sorting, such as σ_g, are known for a site, the other percentiles can be approximated by assuming a log-normal distribution. This can also be used in a piecewise fashion as a means of interpolating between known percentiles. The above method is used in SandCalc-Edit-Bed Material-Derive.

Spot samples of the Quality of the Bottom are marked on UK Admiralty charts. They are marked using symbols to describe the grain size, mineralogy, biological properties, texture and colour of the seabed. Some of the most useful for sediment transport purposes in shelf waters are given in Table 2.

Samples were mainly obtained by the use of tallow on the bottom of the sounding lead in the days of lead-line soundings, and their validity today should be treated with caution.

Mixed sediments can be classified according to the modified Folk scheme (BGS, 1987). In this, the proportions of mud

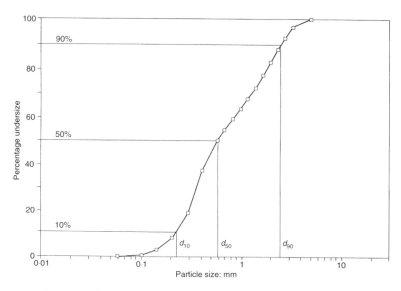

Figure 5. *Example of grain-size distribution*

Table 2. *Symbols used on UK Admiralty charts to describe the sea bed*

S	Sand	f	fine
M	Mud	c	coarse
Cy	Clay	so	soft
G	Gravel	h	hard
Sn	Shingle	sf	stiff
P	Pebbles	bk	broken
St	Stones	sm	small
R	Rock	l	large
Bo	Boulders	Wd	Weed
Sh	Shells	SM/R	Sand and Mud over Rock

($d < 62\cdot5$ μm), sand ($62\cdot5 < d < 2000$ μm) and gravel ($d > 2$ mm) are expressed as (a) the ratio of sand to mud, and (b) the percentage gravel. The predominant type of sediment is given in abbreviated form in capital letters (e.g. S = sand), qualified if necessary with lower-case letters (e.g. m = muddy). Thus a muddy sand is designated mS. The different categories are often presented as a triangular figure, but the rectangular version shown in Figure 6 is easier to use for plotting measured sediment characteristics. The modified Folk classification is used by the British Geological Survey in their marine sediment distribution charts.

The mineralogy of sand and the shape of the grains determine its hydraulic properties. Most sands in European waters are primarily composed of quartz grains. These usually have a density close to 2650 kg m^{-3} and are very roughly spherical (i.e. their major and minor axes do not usually vary by more than a factor of 2). Varying proportions of the whole or broken shells of marine creatures are often present (proportions in the range 20% to 70% shell are common around Britain). Shell has a density of typically about 2400 kg m^{-3}, and both whole shells and fragments are platey, irregular and angular. Other minerals, such as coal with a density of about 1400 kg m^{-3}, may also be present. Grains which have a low density and/or platey shape behave hydraulically like quartz grains of a smaller diameter, and, as a result of hydraulic sorting by the prevailing flows, they are usually found mixed with such grains. As a result, it is common practice for sediment transport calculations to treat the grains in a sample as if they were all represented by the quartz grains on grounds of hydraulic

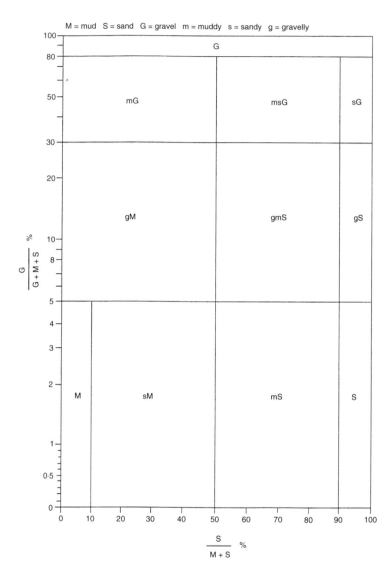

Figure 6. Classification of sediment samples on rectangular Folk diagram

similarity, and that approach will be followed here. However, in some parts of the world sands of a different origin may predominate; for example, coral sands and volcanic sands. In these cases the hydraulic properties, such as settling velocity and threshold of motion, should be measured in the laboratory.

Procedure

1. Sediment samples may be obtained from the study area by one or more of the following techniques
   - grab sample or diver-collected sample of top approximately 20 to 80 mm of seabed
   - box core for undisturbed samples of area approximately 0.2×0.3 m to a depth of approximately 0.3 m
   - gravity core or vibrocore for samples of approximately 75 mm diameter to depths of $1-10$ m
   - borehole log
   - pumped samples of suspended sediment in the water column, taken through nozzles at heights above the bed between about 50 mm and the water surface, filtered on a (typically) 40 μm gauze.

   Other methods of collecting sediment (e.g. bottle samplers and traps, gauze streamers) are also sometimes used in the sea (see Section 2.3).

2. Perform a sieve analysis using the standard technique (BSI, 1967; 1986).

3. Plot a grain-size distribution of '% finer' versus diameter (Figure 5). Read off d_{50}, d_{10} and d_{90} (or d_{16} and d_{84}).

4. Plot ratios of mud: sand: gravel on Figure 6 to determine Folk classification.

2.3. SAND–WATER MIXTURES

Knowledge

Sand–water mixtures occur in ratios ranging from that of a settled immobile sand bed, through slurries and suspensions, to clear water. A wide variety of measures for this ratio are in use, whose usage depends on the type of mixture and the academic background of the user. Definitions of ten of these measures are given in Table 3. Some of these are different names for the same quantity, reducing the number of distinct measures to five. Conversion formulae between these five are given in Table 4.

Table 3. Measures for sand–water mixtures

Quantity	Usage	Definition	Symbol
Volume concentration Packing density Packing fraction	Suspensions (theory) Bed (soil mechanics) Bed (soil mechanics)	$\dfrac{\text{Volume of grains}}{\text{Volume of mixture}}$	C
Mass concentration Dry density	Suspensions (experimental) Bed (cohesive)	$\dfrac{\text{Mass of grains}}{\text{Volume of mixture}}$	C_M
Porosity	Bed (non-cohesive)	$\dfrac{\text{Volume of water}}{\text{Volume of mixture}}$	ε
Voids ratio	Bed (soil mechanics)	$\dfrac{\text{Volume of water}}{\text{Volume of grains}}$	V_R
Suspension density Bulk density Wet density	Suspensions (hydrodynamics) Bed (non-cohesive) Bed (cohesive)	$\dfrac{\text{Mass of mixture}}{\text{Volume of mixture}}$	ρ_B

Table 4. Conversions between measures for sand–water mixtures

From ＼ To	C	C_M	ε	V_R	ρ_B
C	–	C_M/ρ_s	$1-\varepsilon$	$\dfrac{1}{1+V_R}$	$\dfrac{\rho_B-\rho}{\rho_s-\rho}$
C_M	$\rho_s C$	–	$\rho_s(1-\varepsilon)$	$\dfrac{\rho_s}{1+V_R}$	$\dfrac{\rho_s(\rho_B-\rho)}{(\rho_s-\rho)}$
ε	$1-C$	$1-C_M/\rho_s$	–	$\dfrac{V_R}{1+V_R}$	$\dfrac{\rho_s-\rho_B}{\rho_s-\rho}$
V_R	$\dfrac{1-C}{C}$	$\dfrac{\rho_s-C_M}{C_M}$	$\dfrac{\varepsilon}{1-\varepsilon}$	–	$\dfrac{\rho_s-\rho_B}{\rho_B-\rho}$
ρ_B	$\rho(1-C)+\rho_s C$	$\rho+C_M\left(\dfrac{\rho_s-\rho}{\rho_s}\right)$	$\rho\varepsilon+\rho_s(1-\varepsilon)$	$\dfrac{\rho V_R+\rho_s}{1+V_R}$	–

ρ = density of water (typically $1000\,\mathrm{kg\,m^{-3}}$ for fresh water, $1027\,\mathrm{kg\,m^{-3}}$ for sea water).
ρ_s = density of grain material (typically $2650\,\mathrm{kg\,m^{-3}}$ for quartz). See Table 3 for meaning of symbols.
Example: to convert from porosity to mass concentration $C_M = \rho_s(1-\varepsilon)$.

The most fundamental measure is *volume concentration*, and this is usually the simplest to use in theoretical analyses. Experimental results are usually obtained by weighing the sediment, so *mass concentration* is the most appropriate here. Optical, acoustic, resistive and volumetric techniques, however, measure primarily the volume concentration. In discussions of the settled bed, the *porosity* is most often used. When considering buoyancy effects, the *bulk density* is most appropriate.

The porosity of settled sand beds depends on the degree of sorting and the degree of compaction. A compilation of the average porosities measured in natural sand beds is given in Table 5.

The *angle of repose* ϕ_i (or, more precisely, angle of internal friction) is the angle to the horizontal at which grains start to roll on a flat bed of sediment which is gradually tilted from the horizontal. The value of ϕ_i for non-cohesive sediments depends on the shape, sorting and packing of the grains. The concentration (or porosity) of the bed also depends on the same factors, and ϕ_i is found experimentally to increase with the bed concentration (decrease with porosity). Consideration of the forces involved shows that the value of ϕ_i ought not to depend on grain diameter or density, although some old and frequently reproduced plots appear (inexplicably) to show a diameter dependence for given grain shape. They should probably not be trusted.

For the purposes of calculating slope effects on the threshold of motion, or sediment transport rates, a good general purpose value for the angle of repose is $\phi_i = 32°$.

The *angle of final repose* ϕ_r (or angle of residual shearing) is the angle of the final slope after avalanching has ceased. The angles of the lee slopes of ripples, dunes and sandwaves, and the angle of slope of the conical scour hole around a circular vertical pile, are

Table 5. *Porosity of natural sand beds*

	Well-sorted	Average	Well-mixed
Loosely packed	0·46	0·43	0·38
Average	0·42	0·40	0·33
Densely packed	0·40	0·37	0·30

Sources: Terzaghi and Peck (1967), Shepard (1963), Sleath (1984), Dyer (1986); measurements at HR.

given by ϕ_r. The value of ϕ_r is always less than ϕ_i, with a typical value for natural sand in water of about 28°.

Procedure

1. Sand concentration may be measured by one of the following techniques:

   - bottle or trap or gauze streamer samplers
   - pumped water samples
   - optical (or infrared) attenuation or backscatter, or laser diffraction
   - acoustic attenuation or backscatter
   - impact probes (e.g. Sand Transport Probe)
   - electrical resistance techniques (e.g. Coulter Counter, bed conductivity probes)
   - gamma-ray absorption
   - volumetric techniques (e.g. bed porosity determination, volumetric determination of pumped samples).

2. Of the ten alternative measures for sand–water mixtures it is recommended that only volume concentration, mass concentration, porosity and bulk density are used, with a preference for volume concentration for theoretical purposes or measurements by optical, acoustic, resistive and volumetric techniques, and mass concentration for weighed sediment samples, and impact probes.

3. If measurements are obtained from other sources in one of the less recommended measures, convert by using Table 4.

4. The porosity of an undisturbed bed sample can be obtained by measuring its bulk density and converting using Table 4. If an undisturbed bed sample is not available, use Table 5 based on measurement or judgement of the degrees of sorting and packing of the bed sediment. In the absence of any other information, use a default value of $\varepsilon = 0.40$.

5. The angle of repose ϕ_i of a sediment sample can be obtained by measuring the sideslope angle of a conical pile of sediment formed by pouring the sample underwater from a small height. In the absence of a measured value, use a default value of $\phi_i = 32°$.

6. Dredged sediment is commonly measured in hopper tonnes.

One hopper tonne = (bulk density) × (volume of sediment-water mixture). Typical values of the bulk density are 1·7 tonnes/m^{-3} for sand and 1·2 tonnes m^{-3} for silt. Conversions to dry weight of sediment or dredged *in situ* volume (assume *in situ* $\varepsilon = 0·40$) may be made using Table 4.

2.4. PERMEABILITY AND FLUIDISATION

Knowledge

The flow of water through a porous medium, such as a bed of sand, depends on the *permeability* of the medium. If the flow is laminar, as will generally be the case for sand grain sizes smaller than about 1 mm, the relationship of the flow velocity to the pressure gradient that is driving the flow is governed by Darcy's Law. This can be written in two equivalent forms:

$$V_B = \frac{K_p}{\rho\nu}\frac{dp}{dz} \tag{9}$$

$$V_B = K_I I \tag{10}$$

where
$\quad V_B$ = bulk velocity

dp/dz = pressure gradient

$\quad I$ = hydraulic gradient

$\quad \rho$ = density of water

$\quad \nu$ = kinematic viscosity of water

$\quad K_p$ = specific permeability (m^2)

$\quad K_I$ = coefficient of permeability (m s^{-1}).

The bulk velocity V_B is the discharge of water per unit area of bed normal to the flow direction. It is thus a smaller velocity (by a factor ε) than the velocity of the water flowing between the grains. The pressure gradient has been written as dp/dz, appropriate to a vertical flow, but Equation (9) can be adapted to apply to any direction. The hydraulic gradient is a measure of the pressure gradient in which the pressure is expressed as head of water, and hence I is dimensionless. The pressure gradient and hydraulic gradient are related through:

$$\frac{dp}{dz} = g\rho I \tag{11}$$

Hence the two alternative measures of permeability are related through

$$K_{\mathrm{I}} = \frac{gK_{\mathrm{p}}}{\nu} \tag{12}$$

The choice of Equation (9) or (10), and hence the use of K_{p} or K_{I}, is immaterial. Physicists tend to use Equation (9) and soil mechanics/geotechnical engineers tend to use Equation (10). Equation (9) is recommended as the more general expression, as it does not introduce the spurious quantity g that does not have a direct role in permeable flow. In addition, the specific permeability K_{p} is preferred since it depends only on the properties of the porous medium (the sand bed in our case), whereas K_{I} depends on the properties of both the porous medium and the fluid.

For grains larger than about 1 mm the intergranular flow may become turbulent, and the force exerted on the grains is a combination of viscous (linear in velocity) and turbulent (quadratic in velocity) forces. Darcy's Law (in the form of Equation (10)) is then replaced by Forchheimer's equation

$$I = a_{\mathrm{I}} V_{\mathrm{B}} + b_{\mathrm{I}} V_{\mathrm{B}}^2 \tag{13}$$

where $a_{\mathrm{I}} = 1/K_{\mathrm{I}}$ for consistency with Equation (10). Alternatively, Equation (13) can be written in a form analogous to Equation (9):

$$\frac{dp}{dz} = a_{\mathrm{p}} \frac{\rho\nu}{d^2} V_{\mathrm{B}} + b_{\mathrm{p}} \frac{\rho}{d} V_{\mathrm{B}}^2 \tag{14}$$

The coefficients are related via Equations (9) and (11):

$$K_{\mathrm{p}} = \frac{d^2}{a_{\mathrm{p}}}; \quad a_{\mathrm{I}} = \frac{\nu}{gd^2} a_{\mathrm{p}}; \quad b_{\mathrm{I}} = \frac{1}{gd} b_{\mathrm{p}} \tag{15}$$

The dimensionless coefficients a_p and b_p are functions of porosity, and grain shape, packing, orientation, and grading.

Various expressions have been proposed to express the permeability of sand, gravel and rock in terms of grain diameter d and porosity ε. Some of these are summarised and reviewed by Van Gent (1993). They are re-written in terms of the coefficients K_p, a_p and b_p in Table 6, and plotted in Figure 7. Also included is a new version (Soulsby) which follows the same derivation as Equation (103) for the allied problem of the settling velocity of grains through dense suspensions. All the methods give a strong increase of the dimensionless permeability K_p/d^2 with the porosity ε (see Figure 7), but differ by a factor of 2 between methods. The equation of Engelund (1953) is widely used. However, data for sand beds from HR Wallingford and Sleath (1970) show a relatively constant value of K_p/d^2, with a mean of $1\cdot1 \times 10^{-3}$. Thus, if there is any uncertainty about the value of ε to use in a particular study, a safer option is to use the following formula, applicable to grains smaller than $1\cdot2\,\text{mm}$:

$$K_p = 0\cdot0011d^2 \tag{16}$$

Expressions for K_I, a_I and a_p can be derived from Equation (16) using the conversions given previously.

Table 6.   Permeability of sand beds

Source	K_p	a_p	b_p
Ergun (1952)	$\dfrac{\varepsilon^3\,d^2}{150\,(1-\varepsilon)^2}$	$\dfrac{150\,(1-\varepsilon)^2}{\varepsilon^3}$	$\dfrac{1\cdot75\,(1-\varepsilon)}{\varepsilon^3}$
Engelund (1953)	$\dfrac{\varepsilon^2\,d^2}{1000\,(1-\varepsilon)^3}$	$\dfrac{1000\,(1-\varepsilon)^3}{\varepsilon^2}$	$\dfrac{2\cdot8\,(1-\varepsilon)}{\varepsilon^3}$
Koenders (1985)	$\dfrac{\varepsilon^3\,d^2}{290\,(1-\varepsilon)^2}$	$\dfrac{290\,(1-\varepsilon)^2}{\varepsilon^3}$	$\dfrac{1\cdot4}{\varepsilon^5}$
Den Adel (1987)	$\dfrac{\varepsilon^3\,d^2}{160\,(1-\varepsilon)^2}$	$\dfrac{160\,(1-\varepsilon)^2}{\varepsilon^3}$	$\dfrac{2\cdot2}{\varepsilon^2}$
Soulsby	$\dfrac{\varepsilon^{4\cdot7}\,d^2}{19\cdot8\,(1-\varepsilon)}$	$\dfrac{19\cdot8\,(1-\varepsilon)}{\varepsilon^{4\cdot7}}$	$\dfrac{0\cdot956\,(1-\varepsilon)}{\varepsilon^{4\cdot7}}$

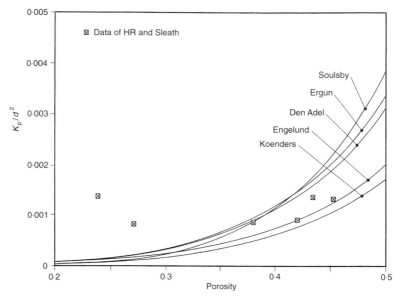

Figure 7. Permeability versus porosity

The permeability also varies with the shape of the grains, their packing and orientation, and the grain size distribution. The expressions given in Table 6 and plotted in Figure 7 are based on the mean values from the range reported by the authors (observed variation is typically within a factor of 2) and for a uniform grain size. For mixed sediments d is replaced by d_{15} (Koenders and Den Adel), or the equivalent sphere diameter (Engelund, 1953).

The above applies to steady flow. In oscillatory flows an additional term proportional to the fluid acceleration is added to the Forchheimer Equation (13). This is important for grain sizes larger than about 10 mm. A full discussion of these effects is given by van Gent (1993). For grains smaller than about 10 mm, Equation (14) can be used in a quasi-steady fashion in oscillatory flows.

Permeability is important in the following applications:

- damping of wave energy, due to viscous pumping of water through the bed (see, for example, Sleath, 1970)
- entrainment of sediment by waves, because pumping of water in and out of the bed will respectively increase and reduce the effective weight of grains

- beach stabilisation by artificial drainage
- transfer of pollutants into and out of the bed
- percolation through gravel beaches
- stability of rubble-mound breakwaters and rock armoured structures

An upward flow of water through a sand bed exerts an upward drag force on the grains, corresponding to the resistance controlling permeable flow. If this force is larger than the buoyant weight of the grains, the bed becomes *fluidised*. The weight of the grains is no longer supported by resting on other grains, but by fluid forces, and the bed behaves like a fluid, so that a heavy object placed on the bed sinks through it.

The minimum vertical pressure gradient necessary to achieve fluidisation is simply that which balances the weight of the grains:

$$\left(\frac{dp}{dz}\right)_{mf} = g(\rho_s - \rho)(1 - \varepsilon) \tag{17}$$

The corresponding minimum fluidisation velocity w_{mf} of upward flowing water necessary to achieve fluidisation is:

$$w_{mf} = \frac{\nu}{d}\{[10\cdot36^2 + 1\cdot049\varepsilon^{4\cdot7}D_*^3]^{1/2} - 10\cdot36\} \tag{18}$$

with ν = kinematic viscosity of water

d = grain diameter

ε = porosity of bed

$$D_* = \left[\frac{g(s-1)}{\nu^2}\right]^{1/3} d$$

g = acceleration due to gravity

s = ratio of densities of grain and water

Equation (18) is compatible with Equation (14) with Soulsby's expressions for a_p and b_p for permeability, Equation (103) for settling of grains in dense suspensions, and Equation (102) for settling of single grains in clear water.

Wen and Yu (1966) gave the following formula, which is similar in form to Equation (18), but does not include a dependence on porosity:

$$w_{mf} = \frac{\nu}{d}\{[33\cdot7^2 + 0\cdot0408D_*^3]^{1/2} - 33\cdot7\} \tag{19}$$

Experiments at HR Wallingford on fluidisation of sand beds showed little dependence on porosity (similar to the result for permeability quoted above), and gave a mean value (applicable to grains smaller than 0·8 mm) of:

$$w_{mf} = 5{\cdot}75 \times 10^{-4}(s-1)\frac{gd_{50}^2}{\nu} \tag{20}$$

Fluidisation is important in areas where there is a natural or artificial upward flow of water through the bed. It can cause catastrophic failure or loss of an object or structure resting on the bed. Fluidisation can also be caused by wave action, although the mechanisms are not fully understood. The quasi-steady application of Equation (17) indicates that wave-induced vertical pressure gradients under non-breaking waves are insufficient to cause fluidisation. None the less, wave-induced fluidisation of the seabed undoubtedly does occur, and may be an important factor in sediment mobility. Explanations include: the higher pressure gradients under breaking waves, or interfering or reflected wave trains; and the pumping effect of a train of waves building up pressure gradients within the bed.

Procedure

1. If a sample of the bed material is available, measure its permeability by using a *permeameter*. This is a vertical cylinder in which the sediment sample is placed, through which a controlled and measured upward flow of water can be produced using a constant-head reservoir. Pressure tappings enable the vertical pressure gradient to be measured.

2. Otherwise, if the porosity is known accurately, use Engelund's formula from Table 6 with Equation (14) to calculate the permeability. If consistency with fluidisation and settling velocity is required, use Soulsby's formula instead. If the porosity is not know accurately, use Equation (16).

3. Fluidisation can also be measured using a permeameter, if a bed sample is available. A small weight placed on the surface of the bed will suddenly sink when the bed fluidises, as the flow speed is gradually increased.

4. Otherwise, calculate the pressure gradient from Equation (17), and the minimum velocity from Equation (18), necessary to produced fluidisation. If porosity is not known, or uncertain, use Equation (20) instead.

Currents

3

3. Currents

3.1. GENERAL

Currents in the sea may be caused by tidal motions, wind-stress, atmospheric pressure gradients, wave-induced forces, river outflows, large-scale quasi-steady water surface slopes and horizontal density gradients associated with oceanic circulations. In the nearshore region, wave-induced (longshore) currents are dominant, whereas further offshore a combination of tidal and meteorological forcing (including storm surges) dominates. Currents both stir up and transport sediments, and hence the sediment transport largely follows the current direction. However, because the sediment transport rate depends non-linearly on the current speed, and also because the effect of wave-stirring is important, the direction of net long-term sediment transport may be very different from the residual current direction.

3.2. CURRENT VELOCITY PROFILE

Knowledge

A current flowing over the seabed experiences friction with the bed which forms a turbulent *boundary layer* typically some metres or tens of metres thick. In shallow water the boundary layer may occupy the entire depth, whereas in deep water it occupies the lower part of the water column and is overlain by water relatively unaffected by friction. Within the boundary layer the current speed increases with height from zero at the seabed to a maximum at or near the water surface, with the most rapid increase with height occurring near the bed. The way in which the current increases with height is known as the current *velocity profile*.

The most commonly used measure of the current speed at a particular time and place is the *depth-averaged current speed*, $\bar{U}$. This is related to the velocity profile $U(z)$ through the definition

$$\bar{U} = \frac{1}{h}\int_0^h U(z)\, dz \qquad (21)$$

where $\bar{U}$ = depth-averaged current speed
          h = water depth
     $U(z)$ = current speed at height z
        z = height above sea bed

The lower limit of the integral has to be changed from 0 to z_0 if the velocity profile tends to zero at a height $z = z_0$ (e.g. for a logarithmic velocity profile, see next paragraph).

Within the bottom few metres above the bed the current velocity U varies with the height z above the bed according to the logarithmic velocity profile

$$U(z) = \frac{u_*}{\kappa}\ln\left(\frac{z}{z_0}\right) \qquad \text{SC (22)}$$

where u_* = friction velocity
     z_0 = bed roughness length
     κ = von Karman's constant = 0·40

The friction velocity u_* is related to the bed shear-stress through the relationship $\tau_0 = \rho u_*^2$ (see Section 1.4).

The range of heights for which Equation (22) is valid is from a few centimetres above the bed up to about 20–30% of the water depth in shallow water (say, up to $z = 2$ to 3 m), or 20–30% of the boundary-layer thickness in deep water (say, up to $z = 20$–30 m).

There has been debate in the past about the value of von Karman's constant to use in the sea, and whether this value is modified when sediment is in suspension. However, present thinking is that the universal value of $\kappa = 0·40$ should be used, and sediment-induced effects on the velocity profile should be treated separately.

Equation (22) applies to steady flow without density stratification over a flat (but possibly rippled) seabed, well away from structures, and outside the surf zone. A tidal flow can to a reasonable approximation be treated as quasi-steady in the bottom 2–3 m, except within about one hour either side of slack water, which is generally not important for sand transport purposes. References to methods for dealing with more complex situations are listed on page 51.

The bed roughness length z_0 experienced by a current depends on the viscosity of the water, the current speed and the dimensions of the physical roughness of the bed. A set of classical experiments by Nikuradse (1933) into the dependence of z_0 on these quantities still forms the basis of the prediction of z_0 in natural and engineering flows.

A good fit to the experimental results of Nikuradse is given by the expression of Christoffersen and Jonsson (1985)

$$z_0 = \frac{k_s}{30}\left[1 - \exp\left(\frac{-u_*k_s}{27\nu}\right)\right] + \frac{\nu}{9u_*} \tag{23a}$$

where ν is the kinematic viscosity of water (see Section 2.1). Equation (23a) is valid for all values of the grain Reynolds number u_*k_s/ν.

A simplified, but less accurate, version of Equation (23a), used for mathematical convenience by Colebrook and White (1937), is

$$z_0 = \frac{k_s}{30} + \frac{\nu}{9u_*} \tag{23b}$$

For *hydrodynamically rough* flow ($u_*k_s/\nu > 70$), Equation (23a) reduces to

$$z_0 = k_s/30 \tag{23c}$$

For *hydrodynamically smooth* flow ($u_*k_s/\nu < 5$), Equation (23a) reduces to

$$z_0 = \nu/(9u_*) \tag{23d}$$

For *transitional* flow ($5 \leq u_*k_s/v \leq 70$), the full Equation (23a) should be used.

Typically, muds and flat fine sands are hydrodynamically smooth or transitional, and coarse sands and gravels are hydrodynamically rough. It is common practice to treat *all* flows over sands as being hydrodynamically rough, since this simplifies the mathematics. This simplifying approximation makes less than 10% error in the calculation of u_*, for all u_* above the threshold of motion (see Section 6.4) of grains larger than 60 μm.

The bed roughness length z_0 for a flat, non-rippled, bed of sand is given in terms of the Nikuradse roughness k_s. Several relationships between k_s and grain size have been proposed, with one of the most widely used being

$$k_s = 2 \cdot 5 d_{50} \tag{24}$$

Sleath (1984, p. 39) and Van Rijn (1993, p. 6.5) list some alternatives to Equation (24). These are given in terms of various grain-size percentiles as: $k_s = 1 \cdot 25 d_{35} \; 1 \cdot 0 d_{65}, \; 2 d_{65}, \; 2 \cdot 3 d_{84}, \; 5 \cdot 1 d_{84},$ $2 d_{90}$ and $3 d_{90}$. There is thus substantial disagreement about the best value to use. Using the grain size distribution given in Example 10.1, the various alternatives give values of k_s in the range 0·219–1·44 mm. This will lead to corresponding uncertainty in derived bed shear-stresses, although this will be much smaller than the uncertainty in k_s, because the shear-stress depends only logarithmically on k_s. Field measurements by Soulsby and Humphery (1990) of a tidal current flowing over an immobile sea bed comprising a flat, featureless mixture of gravel, sand and shell with $d_{10} = 1 \cdot 75$ mm, $d_{50} = 12 \cdot 5$ mm and $d_{90} = 27$ mm found that k_s was equal to $2 \cdot 4 d_{50}$ or $1 \cdot 11 d_{90}$. This lends support to Equation (24).

Combining Equations (23c) and (24) relates z_0 directly to grain size for hydrodynamically rough flows:

$$z_0 = \frac{d_{50}}{12} \qquad \text{SC (25)}$$

The sea-bed often comprises mixed sediments, or non-flat conditions. In such cases, z_0 can be obtained from Table 7, taken from the compilation by Soulsby (1983, Table 5.4) of a large number of field measurements of z_0 over natural sea beds, in which the bottom type is classified in a manner similar to that described in Section 2.2.

The tidal current velocity profile throughout the water column is given with reasonable accuracy by (Soulsby, 1990):

$$U(z) = \frac{\bar{U} \ln(z/z_0)}{\ln(\delta/2z_0) - \delta/2h} \qquad \text{for } z_0 < z < 0 \cdot 5\delta \tag{26a}$$

$$U(z) = \frac{\bar{U} \ln(\delta/2z_0)}{\ln(\delta/2z_0) - \delta/2h} \qquad \text{for } 0 \cdot 5\delta < z < h \tag{26b}$$

where $\bar{U}$ is the depth-averaged current speed, $h =$ water depth

Table 7.   *Mean values of z_0 and C_{100} for different bottom types*

Bottom type	z_0 (mm)	C_{100}
Mud	0·2	0·0022
Mud/sand	0·7	0·0030
Silt/sand	0·05	0·0016
Sand (unrippled)	0·4	0·0026
Sand (rippled)	6	0·0061
Sand/shell	0·3	0·0024
Sand/gravel	0·3	0·0024
Mud/sand/gravel	0·3	0·0024
Gravel	3	0·0047

Sources: see Soulsby (1983, Table 5.4).

and δ is the boundary-layer thickness. Equation (26a) is compatible with Equation (22) near the bed.

The tidal boundary layer (the region in which the frictional and turbulent effects of the seabed are experienced) extends to a height given approximately by (Soulsby, 1983):

$$\delta = 0\cdot0038 \left(\frac{\bar{U}_a \sigma - \bar{U}_b f}{\sigma^2 - f^2} \right) \tag{27}$$

in which

δ = boundary-layer thickness
σ = the radian tidal frequency (e.g. $\sigma = 1\cdot4052 \times 10^{-4}\,\mathrm{rad\,s^{-1}}$ for the M_2 tide)
$f = 1\cdot4544 \times 10^{-4} \times \sin$ (latitude) $\mathrm{rad\,s^{-1}}$ is the Coriolis parameter (e.g. $f = 1\cdot1914 \times 10^{-4}\,\mathrm{rad\,s^{-1}}$ at latitude 55°N)

$\bar{U}_a$ and $\bar{U}_b$ = maximum and minimum values of the depth-averaged current speed through a tidal cycle.

By convention, $\bar{U}_b$ is negative if the tidal current vector rotates clockwise (viewed from above) in the northern hemisphere. For a rectilinear current, such as commonly occurs near a straight coastline, $\bar{U}_b = 0$.

The tidal current velocity throughout the water column is alternatively given by the following empirical formula (Soulsby, 1990)

$$U(z) = \left(\frac{z}{0\cdot32h} \right)^{1/7} \bar{U} \qquad \text{for } 0 < z < 0\cdot5h \qquad \text{SC } (28a)$$

$$U(z) = 1\cdot07\bar{U} \qquad \text{for } 0\cdot5h < z < h \qquad \text{SC } (28b)$$

where $\bar{U}=$ depth-averaged current speed and $h=$ water depth.

Figure 8 shows a comparison of Equation (28) against a wide variety of field measurements in deep and shallow water, slow and fast currents, stratified and unstratified conditions, over flat

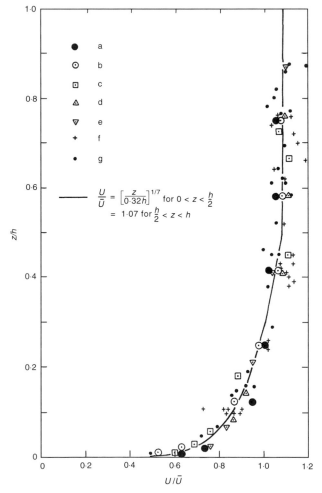

Figure 8. Variation with height of tidal current speed, comparison of Equation (28) with data from: a, b, c — Celtic Sea; d — English Channel; e — Taw Estuary; f — North Sea; g — Celtic and Irish Seas (Reprinted from Soulsby, 1990, in The Sea, 9, Le Mehauté and Hanes (eds), by permission of John Wiley & Sons, Inc. Copyright © 1990 by John Wiley & Sons, Inc.)

beds and sandwaves. The fit is good, with 96% of the data lying within 10% of the curve. Whitehouse (1993) also found good agreement with Equation (28) for detailed velocity profiles in the outer Thames estuary.

However, Equation (28) does not have as strong a basis in physics as Equations (22) and (26).

For flow in more complex conditions, the following references may be useful:

- stratification due to   Haugen (1973)
  temperature   Soulsby (1983, 1990)
- stratification due to   Taylor and Dyer (1977)
  sediment   Soulsby and Wainwright (1987)
- accelerating flow   Soulsby and Dyer (1981)
- fully tidal flow   Prandle (1982a, 1982b)
                          Soulsby (1983, 1990)
- flow over sandwaves Dawson et al. (1983)
                          Johns et al. (1990, 1993)

Procedure

1. To calculate the depth-averaged current speed from measurements of a velocity profile, use a trapezium rule to approximate the integral in Equation (21). Take the velocity at the sea bed ($z = 0$) to be zero, and take the velocity at the water surface ($z = h$) to be equal to the velocity at the highest measuring point. Thus, if the current speeds at heights $z_1, z_2, z_3, \ldots, z_n$ (increasing upwards) are $U_1, U_2, U_3, \ldots, U_n$, then the calculated value of the depth-averaged current speed is

$$\bar{U} = \frac{0{\cdot}5}{h} [U_1 z_1 + (U_1 + U_2)(z_2 - z_1)$$
$$+ (U_2 + U_3)(z_3 - z_2) \ldots + (U_{n-1} + U_n)$$
$$\times (z_n - z_{n-1}) + 2U_n(h - z_n)] \tag{29}$$

Equation (29) is an adequate approximation to U provided that: (a) there are sufficient measurement heights (preferably at least 6); (b) the lowest height is sufficiently close to the bed (not higher than $0{\cdot}2$ m), (c) the highest height is above mid-depth; and (d) the measurements are either simultaneous, or measured over a time-span which is much shorter than the time over which the current is varying (e.g. the whole profile is taken in less than 20 min for tidal currents).

2. To calculate the velocity profile in the bottom few metres, use Equation (22) with z_0 obtained from Equation (25) for a uniform flat immobile sand bed. The calculation of u_* is discussed in Section 3.3.

Example 3.1. Logarithmic velocity profile

Calculate the velocity at a height of 2 m above a flat bed of sand of uniform diameter $d = 200$ μm, if the bed shear-stress is $0.2\,\mathrm{N\,m^{-2}}$ and the water density is $1027\,\mathrm{kg\,m^{-3}}$.

From Equation (1b), $u_* = (\tau_0/\rho)^{1/2} = 0.0140\,\mathrm{m\,s^{-1}}$

From Equation (25), $z_0 = d/12 = 1.67 \times 10^{-5}\,\mathrm{m}$

From Equation (22),

$$U(z = 2\ \mathrm{m}) = \frac{0.0140}{0.40}\ln\left(\frac{2}{1.67 \times 10^{-5}}\right)$$

$$= 0.41\ \mathrm{m\ s^{-1}}$$

3. To calculate the velocity profile throughout the water-column, given the depth-averaged velocity, but without needing to know u_* and τ_0, use Equation (28).

Example 3.2. Power-law velocity profile

Calculate the velocity at a height of 1 m, and the surface velocity, in water of depth 20 m when the depth-averaged velocity is $0.50\,\mathrm{m\,s^{-1}}$.

From Equation (28a),

$$U(z = 1\ \mathrm{m}) = \left(\frac{1}{0.32 \times 20}\right)^{1/7} \times 0.50$$

$$= 0.38\ \mathrm{m\,s^{-1}}$$

From Equation (28b),

$$U(z = 20\ \mathrm{m}) = 1.07 \times 0.50$$

$$= 0.54\ \mathrm{m\,s^{-1}}$$

3.3. CURRENT SKIN-FRICTION SHEAR-STRESS

Knowledge

The bed shear-stress τ_0 (or bottom friction) is the frictional force exerted on unit area of sea bed by the current flowing over it. It is therefore an important quantity for sediment transport purposes, because it represents the flow-induced force acting on sand grains on the bed. Within Section 3.3 it is assumed that the seabed is flat, with no ripples, dunes or sandwaves. In this case, and provided sediment transport is not too intense, the total bed shear-stress τ_0 is equal to the skin-friction contribution τ_{0s}, so for simplicity the subscript 's' is omitted in this Section (also see Section 1.4).

The bed shear-stress is related to the depth-averaged current speed $\bar{U}$ through the drag coefficient C_D by the quadratic friction law

$$\tau_0 = \rho C_D \bar{U}^2 \tag{30}$$

Alternative coefficients used by hydraulic engineers include the Darcy–Weisbach resistance coefficient f, the Chézy coefficient C and Manning's n. When the corresponding laws are written in a form applicable to the sea, these coefficients can be related mathematically to C_D via the relationship

$$C_D = \frac{f}{8} = \frac{g}{C^2} = \frac{gn^2}{h^{1/3}} \tag{31}$$

where h is the water depth and g is the acceleration due to gravity.

The friction (or shear) velocity is an alternative quantity for expressing the friction in velocity units, and is related to τ_0 by

$$u_* = (\tau_0/\rho)^{1/2} \tag{32}$$

The value of C_D is determined by the bed roughness length z_0 (see Section 3.2) and the water depth h.

A simple power law may be used:

$$C_D = \alpha \left(\frac{z_0}{h}\right)^{\beta} \tag{33}$$

Values of α and β which have been proposed are:

Manning–Strickler law: $\alpha = 0.0474$, $\beta = 1/3$
Dawson et al. (1983): $\alpha = 0.0190$, $\beta = 0.208$.

Figure 9 shows a plot of measured values of $(u_*/\bar{U})$ versus

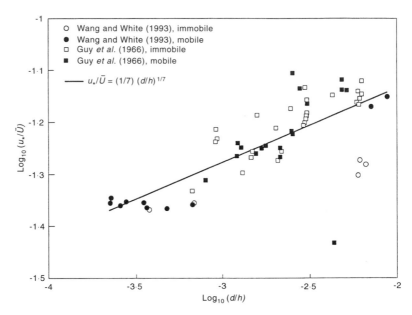

Figure 9. Friction for steady currents on flat immobile and mobile beds

(d_{50}/h) for experiments with steady flows in flumes over flat immobile and mobile beds of sand. There is considerable scatter, but there appears to be no systematic difference between the immobile and mobile beds, and a best power-law fit to all the data yielded the following friction law:

$$\frac{u_*}{\bar{U}} = \frac{1}{7}\left(\frac{d_{50}}{h}\right)^{1/7} \qquad\qquad \text{SC (34)}$$

Substituting $z_0 = d_{50}/12$ (Equation (25)) into Equation (34) yields values which correspond to $\alpha = 0.0415$, $\beta = 2/7$ in Equation (33). These values of the coefficients are recommended for calculating skin-friction (grain-related) bed shear-stresses, in preference to the coefficients of Manning–Strickler and Dawson *et al.* which were derived for other purposes. Equation (34) is compatible with the power-law velocity profile given by Equation (28). Matching Equations (34) and (28) yields an equation for obtaining u_* from a measured velocity $U(z)$ at a height z close to the bed:

$$u_* = 0.121 \left(\frac{d_{50}}{z}\right)^{1/7} U(z) \qquad \text{SC (35)}$$

Alternatively, a logarithmic relationship of the form

$$C_D = \left[\frac{\kappa}{B + \ln(z_0/h)}\right]^2 \qquad (36)$$

can be used.

Commonly, the logarithmic velocity profile (Equation (22)) is assumed to hold throughout the water depth, in which case $\kappa = 0.40$ and $B = 1$ in Equation (36).

This gives the widely used formula:

$$C_D = \left[\frac{0.40}{1 + \ln(z_0/h)}\right] \qquad \text{SC (37)}$$

In deep water, where Equation (26) is taken for the velocity profile through the water column, then $\kappa = 0.40$ and $B = (\delta/2h) - \ln(\delta/2h)$ in Equation (36).

The Colebrook–White formula, Equation (23b), commonly used for rivers, corresponds to $z_0 = (k_s/30) + (\nu/9u_*)$, and $\kappa = 0.405$ and $B = 0.71$ in Equation (36).

The above formulae for C_D are plotted in Figure 10. Apart from the case of a thin boundary layer in deep water ($\delta/h = 0.1$), and the Manning–Strickler law for $z_0/h < 10^{-4}$, the curves are similar to each other.

The choice of method depends on the type of application. The logarithmic form, Equation (36) has the strongest physical justification, but the power-law form is often more convenient for mathematical manipulation, and is more or less equally well supported by data.

Where no information is available, or only a rough estimate is needed, a default value of $C_D = 0.0025$ can be taken.

The bed shear-stress is most closely related to the current near the bed. Consequently, an improvement to Equation (30) for flows with a complex variation in the vertical or in time is given by relating τ_0 to the current speed U_{100} at a height of 1 m above the bed:

$$\tau_0 = \rho C_{100} U_{100}{}^2 \qquad (38)$$

The drag coefficient C_{100} is larger than C_D; values for various bottom types are given in Table 6.

Procedure

1. To calculate the skin friction if the depth-averaged velocity is known, for water at $10\,°C$, $35\,ppt$, $\rho = 1027\,kg\,m^{-3}$, obtain estimates of the following parameters:

   Example 3.3. Skin friction for current

depth-averaged current speed	$\bar{U}$	$1{\cdot}0\,m\,s^{-1}$
grain-diameter of sea bed	d_{50}	$1\,mm$
water depth h		$5\,m$
For power-law velocity profile:		
Use Equation (34) to calculate	u_*	$0{\cdot}0423\,m\,s^{-1}$
Use Equation (32) with $\rho = 1027\,kg\,m^{-3}$ to calculate bed shear-stress	$\tau_0 = \rho u_*^2$	$1{\cdot}84\,N\,m^{-2}$
Alternatively, assume a logarithmic velocity profile holds throughout the depth:		
Calculate	$z_0 = d/12$	$8{\cdot}33 \times 10^{-5}\,m$
Calculate	z_0/h	$1{\cdot}67 \times 10^{-5}\,m$
Use Equation (37)	C_D	$1{\cdot}60 \times 10^{-3}$
Use Equation (30) with $\rho = 1027\,kg\,m^{-3}$	τ_0	$1{\cdot}64\,N\,m^{-2}$

2. For comparison, use of Equation (33) gives

   with Manning–Strickler coefficients $\quad \tau_0 = 1{\cdot}24\,N\,m^{-2}$

   with Dawson *et al.* coefficients $\quad\quad \tau_0 = 1{\cdot}93\,N\,m^{-2}$

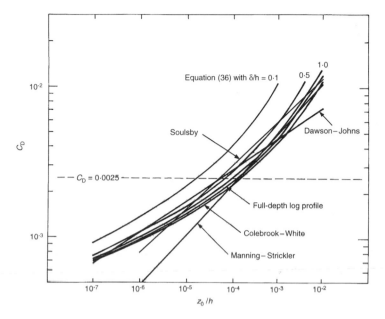

Figure 10. The drag coefficient C_D as a function of the relative roughness. Comparison of various expressions (after Soulsby (1990), in The Sea, 9, Le Mehauté and Hanes (eds), by permission of John Wiley & Sons, Inc. Copyright © 1990 by John Wiley & Sons, Inc.)

The four methods differ by about 50% between the largest and smallest estimates of τ_0.

3.4. CURRENT TOTAL SHEAR-STRESS

Knowledge

In many cases, the bed will not be flat, but will be formed into ripples, dunes or sandwaves. This is the most common condition in the sea outside the surf zone. On non-flat beds with limited sediment transport the *total* bed shear-stress τ_0 is composed of two contributions: the *skin-friction* (or effective shear-stress) component τ_{0s} due to the drag on individual sand grains, and the *form-drag* component τ_{0f} due to the pressure field acting on ripples or larger bedforms:

$$\tau_0 = \tau_{0s} + \tau_{0f} \tag{39}$$

The ratio τ_0/τ_{0s} is typically in the range 2–10 for a rippled bed. Only τ_{0s} is active in moving the sand grains, so it is necessary to have a means of calculating this component when making calculations of the threshold of motion, the bedload transport, or sediment entrainment in the presence of bedforms.

(Also see Section 1.4, Introductory note on bed shear-stresses, and Section 7.4, Friction due to bedforms.)

The method of Einstein (1950) developed for rivers involves the simultaneous solution of two equations:

$$\frac{\bar{U}}{u_{*s}} = 6 + 2{\cdot}5 \ln\left(\frac{\delta_i}{k_s}\right) \tag{40a}$$

$$u_{*s}^{\,2} = g\delta_i I \tag{40b}$$

in which $\tau_{0s} = \rho u_{*s}^{\,2}$, $\bar{U}$ is depth-averaged velocity, g = acceleration due to gravity, I = water surface slope (or hydraulic gradient), $k_s = 2{\cdot}5 d_{50}$ and δ_i is the thickness of the internal boundary layer which grows over a bedform, measured at the crest, with $\delta_i \leq h$. The water surface slope is easily measured in rivers, where levelling techniques can be used over long lengths of river. This method is less useful in tidal flows, because the surface slope I associated with the tidal wave is often not known, and, apart from in very shallow water, the effect of inertia adds an additional important term to Equation (40b).

In very shallow water (say $h < 5\,\text{m}$), the flow is friction-controlled, and if the *total* bed shear-stress τ_0 is known, then I can be calculated from the relationship

$$\tau_0 = \rho g h I \quad \text{for } h < 5 \text{ m} \tag{41}$$

Usually, in the sea, the genuine skin-friction cannot be measured or calculated if bedforms are present. Instead, it is approximated by a *grain-related* bed shear-stress (and friction velocity), which is calculated by using the methods described in Section 3.3 as if there were no bedforms present. This quantity is then used in sediment transport relationships as if it were the true skin-friction value.

The total bed shear-stress τ_0 can be calculated by assigning a total roughness length z_0, which includes both the skin-friction and form-drag components (and sediment-transport component if appropriate; see below). This value can then be used in the formulae in Section 3.3 (e.g. Equation (33) or (37)) to obtain the total drag coefficient, bed shear-stress and friction velocity. Field measurements of near-bottom velocity profiles yield the total friction velocity and the total roughness length. The values of z_0 and C_{100} given in Table 7 correspond to such measured values, and hence can be used for calculating total bed shear-stress.

At very high flow speeds, with intense sheet flow (see Chapter 9), a third component of roughness is found, which arises from the momentum extracted by the flow to move the sand grains. This sediment-transport component of roughness, z_{0t}, is related to the intensity of transport, which in turn is related to the grain-related bed shear-stress τ_{0s}. Wilson (1989a) found from his experiments the relationship

$$z_{0t} = \frac{5\tau_{0s}}{30g(\rho_s - \rho)} \qquad (42)$$

The total bed shear-stress can be calculated by first obtaining the total roughness length z_0 by summing the grain-related, form-drag and sediment-transport components

$$z_0 = z_{0s} + z_{0f} + z_{0t} \qquad (43)$$

in which z_{0s} is given by Equation (25), z_{0t} by Equation (42), and z_{0f} by Equation (90) in Section 7.4. It should be noted, however, that because of the non-linear nature of the equations, the value of τ_0 obtained by calculating z_0 from Equation (43) and then using Equation (37) will be different from the value obtained by computing the three separate components τ_{0s}, τ_{0f} and τ_{0t} by using Equation (37) with z_{0s}, z_{0f} and z_{0t} and then summing. The former procedure is the correct one.

Procedure

1. If measurements of current velocity in the bottom few metres are

available, use Equation (22) to obtain u_* and z_0 and hence $\tau_0 = \rho u_*^2$. Measured values give *total* bed shear-stress.

Example 3.4. Total bed shear-stress from measured profile

Calculate the bed shear-stress acting on an area of seabed above which measurements with current meters at heights of $z = 0{\cdot}1$, $0{\cdot}5$, $1{\cdot}0$ and $2{\cdot}0$ m above the sea bed, averaged over 10 min, were found to give $U(z) = 0{\cdot}20, 0{\cdot}34, 0{\cdot}37$ and $0{\cdot}45\,\mathrm{m\,s^{-1}}$, respectively.

Equation (22) can be written as

$$U(z) = \left(\frac{u_*}{\kappa}\right) \ln z - \left(\frac{u_*}{\kappa}\right) \ln z_0 \tag{44}$$

A linear regression of the form $y = mx + c$ performed with $U(z)$ as y and $\ln z$ as x yields $m = 0{\cdot}08073$, $c = 0{\cdot}3865$, with correlation coefficient $= 0{\cdot}994$. Thus $m = u_*/\kappa$, so

$$u_* = 0{\cdot}40 \times 0{\cdot}08073$$
$$= 0{\cdot}032\ \mathrm{m\,s^{-1}},$$

and $c = m \ln z_0$, so

$$z_0 = \exp(-c/m)$$
$$= \exp(-0{\cdot}3865/0{\cdot}08073)$$
$$= 0{\cdot}0083\ \mathrm{m}.$$

Finally, $\tau_0 = \rho u_*^2$ with $\rho = 1027\,\mathrm{kg\,m^{-3}}$ typically, so
$$\tau_0 = 1{\cdot}05\ \mathrm{N\,m^{-2}}.$$

2. If measurements at only one height are available (preferably at $z = 1$ m), and the bottom type is known, use Equation (38) with C_{100} from Table 7 if $z = 1$ m, or use Equation (22) with z_0 from Table 7 for other heights.

Example 3.5. Total bed shear-stress from U_{100}

Calculate the bed shear-stress acting on an area of rippled sand seabed above which a current meter mounted at a height of 1 m gives an average speed over 10 min of $U_{100} = 0.50 \, \text{m s}^{-1}$.

From Table 7, $C_{100} = 0.0061$ for rippled sand, and typically $\rho = 1027 \, \text{kg m}^{-3}$.

From Equation (32), $\tau_0 = 1027 \times 0.0061 \times 0.50^2 = 1.57 \, \text{N m}^{-2}$.

This method is less accurate than the method given in Example 3.4.

3. For water deeper than about 20 m, the tidal boundary layer may not occupy the full water depth. In this case, to calculate τ_0 it is necessary to use Equation (27) to obtain δ, and then use Equations (26) and (22).

Waves

4

4. Waves

4.1. GENERAL

Waves play a major role in stirring up sediments from the sea bed, as well as giving rise to steady current motions such as longshore currents, undertow, and mass-transport (or streaming) velocities, which transport the sediments. The asymmetry of velocities beneath the crest and the trough of waves is another source of net transport of sediments. Waves may be generated either as a *locally-generated sea* (or wind-sea) due to the effect of local winds blowing over the sea for a certain distance (the fetch), and time (the duration); or as *swell*, which results from distant storms and usually has a longer period and less spread in period and direction than a locally-generated sea. Although most effort has tended to be concentrated on the understanding and effects of locally-generated seas, the swell component, which penetrates easily to the sea bed can play an important role in sediment dynamics.

4.2. WAVE HEIGHT AND PERIOD

Knowledge

The simplest type of water wave is the *monochromatic* (or regular, or single-frequency) wave, having a single value of wave height, H, and wave period, T, each wave being identical to the others. If the wave has a very small height compared to its wavelength (Section 4.3) it approximates well to a sinusoidal variation in surface elevation and orbital velocity (Section 4.4) and its properties are given by linear wave theory (see, for example, Sleath, 1984). Monochromatic waves are often used for simplicity in experiments in laboratory flumes, and in mathematical/physical theoretical derivations involving bed shear-stresses and sediments. Swell waves correspond reasonably well to monochromatic waves.

Natural (irregular, or random) waves in the sea comprise a *spectrum* of wave heights, periods and directions. The frequency spectrum, $S_\eta(\omega)$, gives the distribution of wave energy as a function of the radian frequency $\omega = 2\pi/T$. Measured spectra in the sea can be approximated by various semi-empirical forms. These correspond to the locally-generated waves, and swell waves can be included as an additional contribution at low frequencies. The two most widely used forms are the Pierson–Moskowitz spectrum, which applies to fully-developed waves in deep water, and the JONSWAP spectrum, which has a sharper peak and applies to growing waves in continental-shelf waters (Figure 11). Both spectra can be described by the following equation in terms of the significant wave height H_s (see next paragraph) and the radian frequency ω_p at the peak of the spectrum:

$$S_\eta(\omega) = B\left(\frac{H_s}{4}\right)^2 \frac{\omega_p^4}{\omega^5} \exp\left[\frac{-5}{4}\left(\frac{\omega}{\omega_p}\right)^{-4}\right] \gamma^{\phi(\omega/\omega_p)} \qquad (45a)$$

$$\phi\left(\frac{\omega}{\omega_p}\right) = \exp\left[-\frac{1}{2\beta^2}\left(\frac{\omega}{\omega_p}-1\right)^2\right] \qquad (45b)$$

For Pierson–Moskowitz: $B = 5, \gamma = 1$
For JONSWAP: $B = 3{\cdot}29, \gamma = 3{\cdot}3,$
 $\beta = 0{\cdot}07$ for $\omega \le \omega_p$
 $\beta = 0{\cdot}09$ for $\omega > \omega_p$

The Bretschneider, ITTC and ISSC spectra are other versions which all have the same shape as the Pierson–Moskowitz spectrum. The JONSWAP spectrum is the most appropriate for sediment transport purposes, because it applies to limited depths where the waves 'feel' the bottom, and hence the sediment 'feels' the waves.

Natural waves are most often described only by their *significant wave height*, H_s, and their *mean period*, T_m. These are defined from the zeroth moment m_0 and the second moment m_2 of the spectrum:

$$H_s = 4m_0^{1/2} \qquad (46a)$$

$$T_m = (m_0/m_2)^{1/2} \qquad (46b)$$

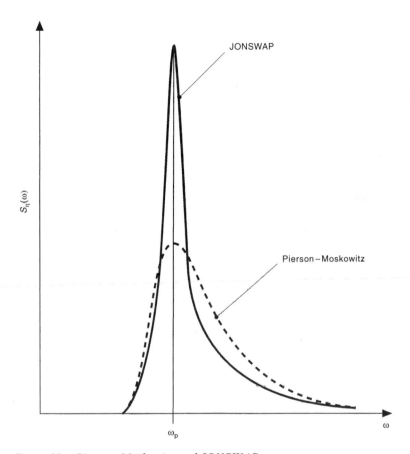

Figure 11. Pierson–Moskowitz and JONSWAP spectra

The moment m_0 is the variance of the elevation of the water surface.

Apart from in the wave breaking zone, these quantities are almost identical with earlier definitions derived from manual wave-counting analysis of wave records, $H_{1/3}(\approx H_s)$ = average height of the highest one-third of the waves, $T_z(\approx T_m)$ = zero up-crossing wave period. The most commonly used quantities are H_s and T_z, and these will be used generally here.

Another useful measure of wave height is the *root-mean-square wave height*, H_{rms}, whose square is a good average measure of the wave energy. Apart from near to breaking, it is related to H_s by the relationship:

$$H_{rms} = H_s/\sqrt{2} \hspace{4cm} \text{SC (47)}$$

Another measure of wave period is the *peak period*, $T_p = 2\pi/\omega_p$, which is the inverse of the frequency at which the peak energy in the wave spectrum occurs. The relationship between T_p and T_z (or T_m) is given theoretically for each spectral shape:

Pierson–Moskowitz: $\quad T_z = 0.710 T_p \hspace{3cm} (48a)$

JONSWAP: $\hspace{1.5cm} T_z = 0.781 T_p \hspace{2.6cm} \text{SC } (48b)$

The directional spreading of waves, which produces *short-crested* waves, is most simply accounted for by multiplying the non-directional spectrum (Equation (45a)) by a spreading function $A \cos^{2n} \alpha$, where α is the direction relative to the mean direction of wave propagation. A commonly used value of the power n is $n = 1$. The coefficient A is chosen so that the energy in all directions sums to the same total as the non-directional spectrum. More sophisticated directional spectra are described by Tucker (1991).

Wave heights and periods determined from a series of 3-hourly recordings can be presented in an $H_s - T_z$ scatter diagram (see Figure 12), which gives the fraction of wave conditions found within each of a number of pre-defined classes of H_s and T_z. The series should be sufficiently long (typically one or more years) that some extreme storm conditions are encountered. The distribution of H_s and direction can similarly be presented as a scatter diagram. These scatter diagrams represent the *wave climate*. They are important for determining long-term patterns of sediment transport, taking account of the relative frequencies of occurrence of calm, moderate and storm conditions.

Sometimes only the value of H_s is known for a particular site, and it is necessary to estimate the corresponding wave period. Figure 12 shows that T_z increases broadly with H_s. An equation which approximates the relationship of the most frequently occurring value of T_z for a given value of H_s, derived by analysing $H_s - T_z$ scatter diagrams from a number of shallow water sites, is:

$$T_z = 11 \left(\frac{H_s}{g} \right)^{1/2} \hspace{4cm} \text{SC (49)}$$

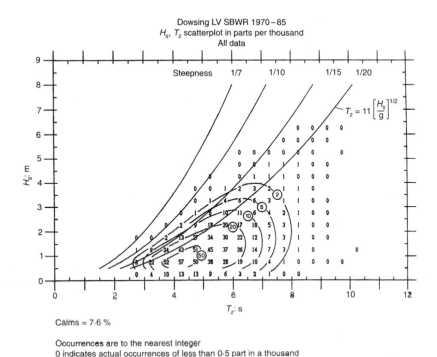

Figure 12. *Scatter diagram indicating joint distribution of H_s and T_s (reprinted from Draper (1991), by permission of Her Majesty's Stationery Office)*

This relationship corresponds approximately to a deep-water wave steepness of 1/20, as shown in Figure 12.

Many of the theoretical expressions for bed shear-stress and sediment dynamics are given in terms of monochromatic wave parameters, usually the bottom orbital velocity amplitude U_w and the period T. The question arises as to which values of U_w and T should be used to represent the full spectrum in a real sea condition.

The simplest approximation is to use a monochromatic wave of height $H = H_{rms}$ and period $T = T_p$. On energy arguments, this matches both the energy density of the waves and the frequency at which the energy is greatest.

Ockenden and Soulsby (1994) showed that the mean bedload transport of sediment by a spectrum of waves plus a current can be simulated to within 20% in magnitude and 10° in direction by an equivalent monochromatic wave having $U_w = \sqrt{2}U_{rms}$, $T = T_p$, and travelling along the mean direction of the (directional) spectrum, where U_{rms} is the standard deviation of orbital velocity for the spectrum. This result applies to a wide range of wave, current and sediment conditions tested, with the exception of cases with very small current speeds ($< 0.2\,\mathrm{m\,s^{-1}}$). The shape of the spectrum (JONSWAP or Pierson–Moskowitz) and the width of directional spreading had very little effect on the mean sediment transport.

A similar result, namely that the equivalent monochromatic wave for calculating sediment transport should have $H = H_{rms}$ and $T =$ significant wave period, was presented by Fredsøe and Deigaard (1992).

Some of the relationships are implemented in SandCalc under Edit-Waves-Derive.

A standard set of notation for describing wave parameters has been recommended by IAHR/PIANC (1986), and has been followed here as far as possible. A thorough account of the measurement, analysis and interpretation of ocean waves is given by Tucker (1991).

4.3. WAVELENGTH

Knowledge

The wavelength L of a water wave is larger for waves of longer period T. The wavelength also becomes shorter as the water depth h decreases. These two effects are expressed by the *dispersion relation*, usually given in terms of the wave number $k = 2\pi/L$, and the radian frequency, $\omega = 2\pi/T$:

$$\omega^2 = gk\,\tanh(kh) \hspace{4cm} \text{SC (50)}$$

where g is the acceleration due to gravity, and tanh is the hyperbolic tangent.

Procedure

1. If L (and hence k) is known, it is easy to calculate ω (and hence T) from Equation (50). However, usually it is the wave period T

which is known, and it is not so straightforward to obtain k (or L) from Equation (50). Writing $\xi = \omega^2 h/g$, and $\eta = kh$, Equation (50) becomes

$$\xi = \eta \tanh \eta \qquad (51)$$

Figure 13 shows a plot of η versus ξ, from which η (and hence λ) can be obtained for a given ξ (and hence T).

Equation (51) can be solved for η given ξ using Newton–Raphson iteration on a computer. This method is used by SandCalc under Waves-Wavelength-Dispersion Relation.

For $\xi < 0.1$ (shallow water, long period waves), Equation (51) becomes approximately $\eta = \xi^{1/2}$, while for $\xi > 3$ (deep water, short period waves), Equation (51) becomes approximately $\eta = \xi$. These convert to $L = (gh)^{1/2}T$ for shallow water, and $L = gT^2/(2\pi)$ for deep water.

2. A simple approximation by G. Gilbert (personal communication) to Equation (51), accurate to 0.75%, which is useful for mathematical manipulations, is given by

$$\eta = \xi^{1/2}(1 + 0.2\xi) \qquad \text{for } \xi \le 1 \qquad (52a)$$

$$\eta = \xi[1 + 0.2 \exp(2 - 2\xi)] \qquad \text{for } \xi > 1 \qquad (52b)$$

Example 4.1. Wavelength

Calculate the wavelength L of a wave of period T	8 s
in water of depth h	10 m
Calculate $\omega = 2\pi/T$	0.785 rad s^{-1}
Calculate $\xi = \omega^2 h/g$	0.629
Calculate η from Equation (52a)	0.893
Calculate $k = \eta/h$	0.0893 m^{-1}
Calculate $L = 2\pi/k$	70.4 m

Alternatively, with reduced accuracy, use Figure 13 to obtain η from ξ.

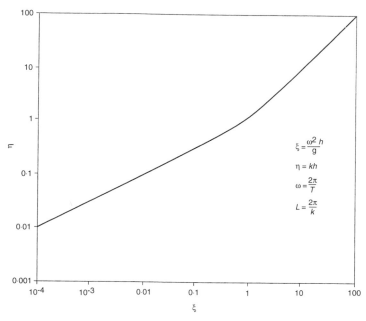

Figure 13. Wave dispersion relation

4.4. WAVE ORBITAL VELOCITY

Knowledge

Waves in sufficiently shallow water produce an oscillatory velocity at the sea bed, which acts on the sediments. 'Sufficiently shallow' in this context means approximately

$$h < 0{\cdot}1gT^2 \tag{53a}$$

or, alternatively

$$h < 10H_s \tag{53b}$$

where h = water depth, H_s = significant wave height, T = wave period, g = acceleration due to gravity.

In practice, wave effects from extreme storms will reach the sea bed over most of the continental shelf.

In this Section the waves are assumed to be non-breaking. Wave breaking and the surf zone are discussed in Section 4.7.

The amplitude U_w of the wave orbital velocity just above the bed due to a monochromatic (single frequency) wave of height H and period T in water of depth h is

$$U_{\mathrm{w}} = \frac{\pi H}{T \sinh(kh)} \qquad \text{SC (54)}$$

where sinh is the hyperbolic sine, $k = 2\pi/L$ is the wave number, and L is the wavelength. As pointed out in Section 4.3, it is not easy to calculate k. Figure 14 gives a curve marked 'Monochromatic' based on Equation (54), from which U_{w} can be calculated directly from the input parameters H, T, h and g, via the quantity $T_{\mathrm{n}} = (h/g^{1/2})$.

In the sea, a spectrum of waves of different heights, periods and directions will be present (see Section 4.2). This produces a random time series of orbital velocity at the sea bed, which can be characterised by its standard deviation U_{rms}. The waves are usually characterised by their significant height H_{s} and zero-crossing period T_{z}. One of the most widely used spectra is the JONSWAP spectrum (see Section 4.2), based on a large number of wave measurements in the North Sea. Figure 14 gives a curve marked 'JONSWAP', from which U_{rms} can be calculated from the input parameters H_{s}, T_{z}, h and g. This curve was derived by applying Equation (54) frequency by frequency through the JONSWAP spectrum, and integrating the results to obtain U_{rms}. Soulsby (1987a) gives a similar curve in terms of the period T_{p} at the peak of the spectrum, and also a second curve corresponding to the Pierson–Moskowitz spectrum.

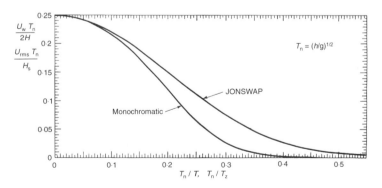

Figure 14. Bottom velocity for monochromatic waves ($U_w T_n/2H$ versus T_n/T) and random waves ($U_{rms} T_n/H_s$ versus T_n/T_z)

Algebraic approximations to the curves in Figure 14 were given by Soulsby and Smallman (1986), and are used in SandCalc under Waves-Orbital Velocity–Spectrum.

The value of U_w given by Equation (54) applies to waves whose steepness (= height/wavelength) is very small, in which case the magnitude of U_w is the same beneath the crest and the trough of the wave. The orbital velocity beneath the wave crest is in the same direction as the wave is travelling, and under the wave trough it is in the opposite direction. In practice, the waves of most interest for sand transport will have a larger steepness. In this case, the maximum velocity under the crest, U_{wc}, is still given reasonably accurately by Equation (54) and Figure 14, but the velocity U_{wt} under the trough is smaller by a factor of up to 1·5 or even 2.

A variety of non-linear wave theories have been devised to deal with steep waves in deep or shallow water. These include:

- Stokes (2nd–5th)-order solutions, valid for water deeper than about $0·01gT^2$
- cnoidal theories, for shallower water of depth between about $0·003gT^2$ and $0·016gT^2$
- stream-function theory, for depths in the range $0·006gT^2$ to $0·016gT^2$
- vocoidal and covocoidal theories, for shallow water including sloping beds

These and other wave theories are described in more detail by Sleath (1984), Tucker (1991), Barltrop (1990), Soulsby et al. (1993), and Kirkgöz (1986), including guidance as to which theory is best for which conditions. For many sediment transport applications, the use of Stokes' 2nd-order theory is adequate, which gives:

$$U_{wc} = U_w\left[1 + \frac{3kh}{8\sinh^3(kh)}\frac{H}{h}\right] \qquad (55a)$$

$$U_{wt} = U_w\left[1 - \frac{3kh}{8\sinh^3(kh)}\frac{H}{h}\right] \qquad (55b)$$

with U_w given by Equation (54). Equation (55a) will tend to overestimate U_{wc}, however. Alternatively, use the method of Isobe and Horikawa (1982),

$$U_{wc} = U_w \qquad (56a)$$

$$U_{wt} = U_w[1 - r_2 \exp(-r_3 h/L_0)] \qquad (56b)$$

with $r_2 = 3\cdot2(H_0/L_0)^{0\cdot65}$ and $r_3 = -27\log_{10}(H_0/L_0) - 17$, where H_0 and $L_0 = gT^2/(2\pi)$ are the deep-water wave-height and wavelength.

The asymmetry between velocities beneath crest and trough is important for sediment transport, and tends to drive sediment onshore.

Procedure

1. To calculate the bottom orbital velocity amplitude for monochromatic (e.g. laboratory flume) waves, use the 'monochromatic' curve in Figure 14.

Example 4.2. Orbital velocity (monochromatic)

Obtain values of:

wave height	H	0·05 m
period	T	1·0 s
water depth	h	0·20 m
Calculate $(h/g)^{1/2} = T_n$		0·143 s
Calculate T_n/T		0·143
Use 'monochromatic' curve in Figure 14 to obtain $U_w T_n/(2H)$		0·183
Calculate $0\cdot183 \times (2H)/T_n$ to obtain bottom orbital velocity amplitude	U_w	0·128 m s^{-1}

2. To calculate the standard deviation of the bottom orbital velocity beneath a JONSWAP spectrum of waves, use the 'JONSWAP' curve in Figure 14.

Example 4.3. Orbital velocity (spectrum)

Obtain values of:

significant wave height	H_s	3·0 m
zero crossing period	T_z	8 s

water depth	h	10 m
Calculate $(h/g)^{1/2} = T_n$		1·01 s
Calculate T_n/T_z		0·126
Use 'JONSWAP' curve in Figure 14 to obtain $U_{rms}T_n/H_s$		0·205
Calculate 0·205 × H_s/T_n to obtain RMS bottom orbital velocity	U_{rms}	0·609 m s^{-1}

4.5.   WAVE SKIN-FRICTION SHEAR-STRESS

Knowledge

Frictional effects near the bed produce an oscillatory boundary layer within which the wave orbital velocity amplitude increases rapidly with height from zero at the bed to the value U_w at the top of the boundary layer. For a smooth bed and relatively small orbital velocities the boundary layer may be laminar, but more often in cases where sediment is in motion it will be turbulent. In the absence of a current the turbulence is confined within the boundary layer, which for waves is only a few millimetres or centimetres thick, in contrast to the boundary layer of a steady current which may be metres or tens of metres thick. This has the effect of producing a much larger velocity shear in the wave boundary layer, which in turn causes the bed shear-stress produced by a wave with orbital velocity amplitude U_w to be many times larger than that produced by a steady current with an equal depth-averaged speed $\bar{U}$.

As with currents (see Section 3.3), the most important hydrodynamic property of waves for sediment transport purposes is the bed shear-stress they produce. This is oscillatory in the case of waves, having an amplitude τ_w. It is usually obtained from the bottom orbital velocity U_w of the waves via the *wave friction factor* f_w, defined by

$$\tau_w = \tfrac{1}{2}\rho f_w U_w^2 \qquad (57)$$

In Section 4.5 it is assumed that the bed is flat, with no ripples. This is generally the case in the surf zone, where the flow is too intense for ripples to exist. The total bed shear-stress amplitude, τ_w, is equal to the skin-friction contribution τ_{ws}, in this case, and the subscript 's' will be omitted (see Section 1.4).

The wave friction factor is dependent on whether the flow is laminar, smooth turbulent, or rough turbulent, which in turn depends on the wave Reynolds number R_w and the relative roughness r:

$$R_w = \frac{U_w A}{\nu} \tag{58a}$$

$$r = \frac{A}{k_s} \tag{58b}$$

where
U_w = bottom orbital velocity amplitude
$A = U_w T / 2\pi$ = semi-orbital excursion
T = wave period
ν = kinematic viscosity
k_s = Nikuradse equivalent sand grain roughness

Myrhaug (1989) gives an implicit relationship for f_w, which makes use of Equation (23a), and is valid in smooth, transitional and rough turbulent flows:

$$\frac{0.32}{f_w} = \left\{ \ln(6.36 r f_w^{1/2}) - \ln\left[1 - \exp\left(-0.0262 \frac{R_w f_w^{1/2}}{r}\right)\right]\right.$$
$$\left. + \left(\frac{4.71 r}{R_w f_w^{1/2}}\right)^2 \right\} + 1.64 \tag{59}$$

For rough turbulent flow a number of formulae have been proposed for the rough bed friction factor f_{wr}:
Swart (1974):

$$f_{wr} = 0.3 \qquad \text{for } r \leq 1.57 \qquad \text{SC (60a)}$$

$$f_{wr} = 0.00251 \exp(5.21 r^{-0.19}) \qquad \text{for } r > 1.57 \qquad \text{SC (60b)}$$

Nielsen (1992):

$$f_{wr} = \exp(5.5 r^{-0.2} - 6.3) \qquad \text{for all } r \qquad \text{SC (61)}$$

Soulsby:

$$f_{wr} = 1 \cdot 39 \left(\frac{A}{z_0} \right)^{-0 \cdot 52} \qquad \text{for all } r \qquad (62a)$$

which can also be written using $z_0 = k_s/30$ as

$$f_{wr} = 0 \cdot 237 r^{-0 \cdot 52} \qquad \text{for all } r \qquad \text{SC } (62b)$$

Equation (62a) was obtained by fitting the two coefficients to the 44 measured values of f_w shown in Figure 15. The data are taken from seven sources, as detailed in Figure 9 of Soulsby et al. (1993). Table 8 shows an error analysis of the fit of Equations (59)–(62) to this data set, giving the percentage of predictions lying within 10%, 20% and 50% of the observations.

The new formula, Equation (62), performs best (as may be expected, since it was fitted to this data). Myrhaug's formula, however, has the advantage of being applicable also to smooth and transitional flow. The data and Equation (62) are shown in Figure 15.

The smooth bed friction factor f_{ws} can be calculated from

$$f_{ws} = B R_w^{-N} \qquad (63)$$

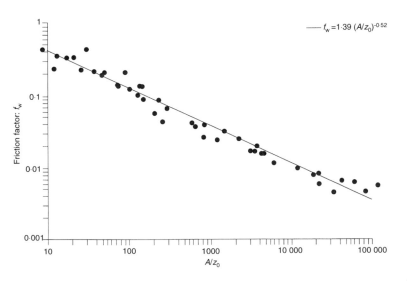

Figure 15. Variation of wave friction factor f_w with relative bed orbital excursion A/z_0

Table 8. Error analysis of fit of Equations (59)–(62)

Formula	Equation	10%	20%	50%
Myrhaug	59	9	16	36
Swart	60a, 60b	5	18	39
Nielsen	61	15	23	37
Soulsby	62a, 62b	19	27	42

where

$$B = 2, \qquad N = 0\cdot5 \quad \text{for } R_\text{w} \leq 5 \times 10^5 \text{ (laminar)}$$

$$B = 0\cdot0521, \quad N = 0\cdot187 \text{ for } R_\text{w} > 5 \times 10^5$$

$$\text{(smooth turbulent)}$$

Alternative values of the coefficients for smooth turbulent flow are $B = 0\cdot035$, $N = 0\cdot16$ (Fredsøe and Deigaard, 1992); $B = 0\cdot0450$, $N = 0\cdot175$ (Myrhaug, 1995).

Procedure

1. For monochromatic waves, calculate U_w from wave height H and period T using Figure 14 (see Example 4.2).

2. For a natural spectrum of waves, calculate U_rms from significant height H_s and zero-crossing period T_z, using Figure 14 (see Example 4.3).
   Take $U_\text{w} = \sqrt{2}U_\text{rms}$, and $T = T_\text{p} = 1\cdot281T_\text{z}$, to represent the amplitude of a monochromatic wave having the same velocity variance as the full spectrum.

3. For flat uniform sand calculate $z_0 = d_{50}/12$.
   For rippled or non-uniform sediments obtain z_0 from Table 7.

4. Calculate $A = U_\text{w}T/2\pi$, and f_wr from Equation (62a).

5. Calculate R_w from Equation (58a) and f_ws from Equation (63) with appropriate coefficients.

6. Take f_w as the greatest of f_wr and f_ws.

7. Calculate τ_w from Equation (57).

Example 4.4. Skin friction for waves

For the wave conditions in Example 4.3, we had $H_s = 3\,\text{m}$, $T_z = 8\,\text{s}$, $h = 10\,\text{m}$, which gave

$$U_{\text{rms}} = 0.609\,\text{m s}^{-1}.$$

The equivalent monochromatic wave has

$$U_w = \sqrt{2} \times 0.609 = 0.861\,\text{m s}^{-1},$$

and $T = 1.281 \times 8 = 10.2\,\text{s}$.

If the bed is a smooth sand with $d_{50} = 480\,\mu\text{m}$, then from Equation (25),

$$z_0 = (480 \times 10^{-6}/12) = 4.0 \times 10^{-5}\,\text{m}.$$

Then $A = 0.861 \times 10.2/2\pi = 1.40\,\text{m}$, and Equation (62a) gives

$$f_{\text{wr}} = 1.39(1.40/4.0 \times 10^{-5})^{-0.52} = 0.00603.$$

Also, if $v = 1.36 \times 10^{-6}\,\text{m}^2\,\text{s}^{-1}$ and $\rho = 1027\,\text{kg m}^{-3}$, then

$$R_w = 0.861 \times 1.40/(1.36 \times 10^{-6}) = 8.85 \times 10^5.$$

Thus $R_w > 5 \times 10^5$, so take smooth turbulent coefficients

$$B = 0.0521, \quad N = 0.187$$

and Equation (63) gives

$$f_{\text{ws}} = 0.0521 \times (8.85 \times 10^5)^{-0.187} = 0.00402.$$

The flow is rough turbulent, because $f_{\text{wr}} > f_{\text{ws}}$.

Then $f_w = \max(f_{\text{wr}}, f_{\text{ws}}) = 0.00603$ and

$$\tau_w = \tfrac{1}{2} \times 1027 \times 0.00603 \times 0.861^2 = 2.30\,\text{N m}^{-2}.$$

4.6. WAVE TOTAL SHEAR-STRESS

Knowledge

In most areas in shallow seas, apart from the surf zone, the bed is generally formed into ripples. These may be current-generated (see Section 7.2) or wave-generated (see Section 7.3), but it is assumed in this Section that they are wave-generated. There may also be dunes or sandwaves present. As in the case of currents (see Section 3.4), on non-flat beds with limited sediment transport, the *total* wave-induced bed shear-stress τ_w is composed of a skin-friction component τ_{ws} and a *form-drag* component τ_{wf}:

$$\tau_w = \tau_{ws} + \tau_{wf} \tag{64}$$

(Also see Section 1.4, Introductory note on bed shear-stresses.)

Various methods of calculating the dimensions and effective roughness of wave-generated ripples are given in Sections 7.3 and 7.4. They enable a value of z_0 to be calculated, which may contain a term corresponding to the effects of intense sediment transport, as well as terms for the skin-friction (or grain-related) roughness and the ripple form-drag roughness.

The difficulty of calculating the true skin-friction component of the current bed shear-stress described in Section 3.4 applies equally to waves. For the same reasons, it is usual to approximate this by a grain-related bed shear-stress (which we will also denote by τ_{ws}, and call the skin-friction bed shear-stress) obtained by using a value of $z_0 = d_{50}/12$ in the friction calculations. The rationale behind doing this is that the quantity τ_{ws} can be calculated relatively easily and unambiguously, and can serve as an independent variable to which dependent variables such as ripple dimensions or sediment reference concentrations can be related.

Procedure

1. The calculation of the total wave bed shear-stress proceeds in the same way as for the flat-bed case described in Section 4.5, but with the value of z_0 being one which includes the form-drag and transport components.

2. SandCalc provides the following methods under Hydrodynamics – Waves – Total shear-stress:

   - Raudkivi: calculates Δ_r and λ_r by the method of Nielsen (1992) (Equation (89)), z_0 by the method of Raudkivi (1988) (Equations (90) and (93)), f_{wr} by the method of Swart (1974) (Equation (60)), and τ_w by Equation (57).

- Nielsen: same as Raudkivi, but calculates z_0 by method of Nielsen (1992) (Equations (90) and (92)).
- Grant and Madsen: calculates Δ_r, λ_r and f_{wr} by method of Grant and Madsen (1982) (Equations (88), (90) and (91)), f_{wr} by method of Soulsby (Equation (62)) and τ_w by Equation (57).

3. An illustration of the procedure for calculating the total wave bed shear-stress is given in Example 7.5.

4.7.   WAVE BREAKING

Knowledge

The most intense sediment transport in the coastal zone is often found beneath breaking waves, either in the surf zone on a beach or over a sandbank. Because surf-zone processes are complicated and not well understood it is common practice to use results from non-breaking waves here, although this is not strictly justified.

Some support for this approach is provided by the experiments of Deigaard *et al.* (1991), who showed that the bed shear-stresses in the surf zone were not *on average* very different from those of unbroken offshore waves, but they exhibited much greater wave-to-wave variability, so that occasional very large values could occur. However, the neglect of wave breaking could lead to an underestimate of suspended sediment concentrations in the upper half of the water column in the surf zone, where turbulence derived from the breaking process is important.

A common alternative approach is that surf-zone processes are heavily parameterised and treated largely empirically.

The height H at which a monochromatic wave of wavelength L breaks in water of constant depth h is given by the Miche criterion:

$$\frac{H}{L} = 0 \cdot 142 \tanh\left(\frac{5 \cdot 5h}{L}\right) \qquad \text{SC } (65a)$$

In the deep-water limit, Equation (65a) reduces to

$$H = 0 \cdot 142L \qquad (65b)$$

and, in the shallow-water limit, Equation (65a) reduces to

$$H = 0 \cdot 78h \qquad (65c)$$

On sloping beaches a greater wave height can be obtained, increasing with beach slope up to about 20% higher than that given by Equation (65a) for a beach slope of 1:10.

Random waves of significant height H_s break on a beach in water of depth h given to a first approximation by

$$H_s = \gamma h \tag{66}$$

where the coefficient γ takes the value 0·55 for a horizontal bed, and increases with beach slope and with wave period for sloping beds (HRS, 1980).

An alternative criterion for random waves proposed by Battjes and Stive (1985), based on analysis of a large number of laboratory and field measurements, gave the formula:

$$\frac{H_{rms}}{h} = 0\cdot5 + 0\cdot4 \tanh\left(\frac{33H_0}{L_0}\right) \tag{67}$$

where H_0 is the offshore value of the root-mean-square wave height H_{rms}, and L_0 is the wavelength at the peak of the offshore wave spectrum. No dependence on beach slope was found in their analysis.

A review of wave breaking is given by Southgate (1995a), and a review of surf-zone processes generally is given by Kraus and Horikawa (1990).

Procedure

1. Determine if waves are breaking by use of criterion (65a) for monochromatic waves, or criterion (66) or (67) for random waves.

2. To calculate bed shear-stresses, treat as non-breaking waves.

Combined waves and currents

5

5. Combined waves and currents

5.1. GENERAL

In most parts of coastal and shelf seas, both waves and currents play important roles in sediment dynamics. The treatment in this case is complicated by the fact that the waves and currents interact with each other hydrodynamically, so that their combined behaviour is not simply a linear sum of their separate behaviours. Ways in which waves and currents interact include:

- modification of the phase speed and wavelength of waves by the current, leading to refraction of the waves
- interaction of the wave and current boundary layers, leading to enhancement of both the steady and oscillatory components of the bed shear-stress
- generation of currents by the waves, including longshore currents, undertow, and mass transport (streaming) currents.

The first two of these mechanisms are dealt with here.

5.2. WAVELENGTH

Knowledge

Waves propagating in the presence of a current are modified when viewed by a stationary observer, because the wave equations apply in a frame of reference moving with the current speed. The Doppler effect causes waves of a given wavelength to appear to have a shorter period if they are carried by a current towards the stationary observer. For the more usual case where the wave period is fixed (to conserve the total number of waves), the wavelength decreases when the waves encounter an opposing current, and the wave height increases (to conserve the rate of energy transmission). The reverse happens with a current following the waves. The component of current perpendicular to the direction of wave travel has no effect on the waves.

The dispersion relation for waves of absolute radian frequency ω and wavenumber k (see Section 3.2) in the presence of a depth-averaged current of speed $\bar{U}$ at an angle ϕ to the direction of wave travel is

$$(\omega - \bar{U}k \cos \phi)^2 = gk \tanh (kh) \qquad \text{SC (68)}$$

where g is the acceleration due to gravity and h is the water depth. The quantity $(\omega - \bar{U}k \cos \phi)$ is called the *relative* (radian) wave frequency. The absolute frequency is that seen by a stationary observer, whereas the relative frequency is that seen by an observer travelling with the current.

For zero current ($\bar{U} = 0$), Equation (68) reduces to Equation (50). The direction ϕ is such that $\phi = 0°$ when the current travels in the same direction as the wave, and $\phi = 180°$ for a current in the opposite direction to that of the wave travel. Angles other than $\phi = 0°$, 180° and ±90° result in refraction of the wave by the current. For a sufficiently large opposing current ($\bar{U} > \omega/k$) the wave cannot propagate.

Procedure

1. The dispersion relation, Equation (68), can be solved computationally for k, and hence wavelength $L = 2\pi/k$, as a function of $\omega = 2\pi/T$, $\bar{U}$ and ϕ either by Newton–Raphson iteration, or by the computationally efficient method described by Southgate (1988).

2. Calculations of wave propagation, transformation, wavelength and orbital velocity require the current-modified dispersion relation, Equation (68), to be used. However, in calculations of boundary-layer behaviour the absolute (not relative) wave frequency ω should be used.

5.3.   BED SHEAR-STRESSES

Knowledge

The bed shear-stresses beneath combined waves and currents are enhanced beyond the values which would result from a simple

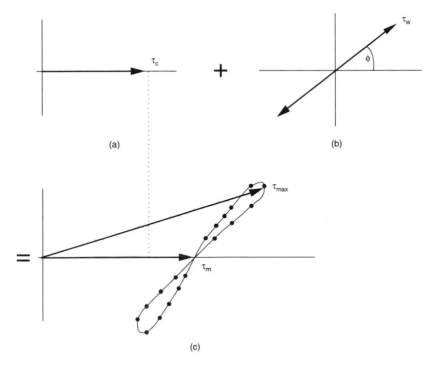

*Figure 16. Schematic diagram of non-linear interaction of wave and current bed shear-stresses (reprinted from Soulsby et al. (1993), Coastal Engineering, **21**, 41–69, by permission of Elsevier Science Publishers, BV)*

linear addition of the wave-alone and current-alone stresses (Figure 16). This occurs because of a non-linear interaction between the wave and current boundary layers. More than 20 different theories and models have been proposed to describe this process. A comparison between predictions of the mean (τ_m) and maximum (τ_{max}) bed shear-stress during a wave cycle by eight of these models is shown in Figure 17. Differences between the models of 30%–40% are commonly found, and differences of up to a factor of 3 occur for strongly wave-dominated conditions. An algebraic approximation (accurate to ±5% in most cases) to the models was derived by Soulsby et al. (1993), and is given below.

A test of the parameterised models against a large data set of 61 laboratory values and 70 field values of τ_m was performed by Soulsby (1995a). Various criteria for goodness of fit were used, and no one model performed best according to all the criteria.

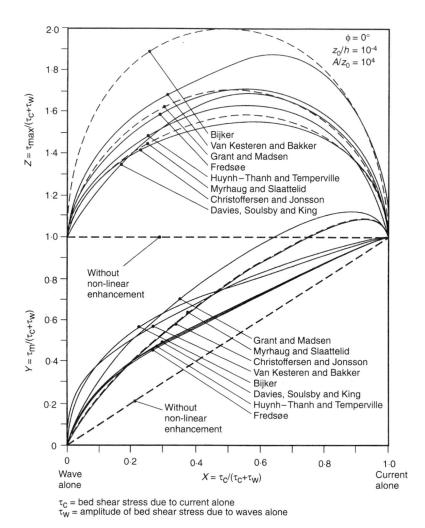

τ_c = bed shear stress due to current alone
τ_w = amplitude of bed shear stress due to waves alone

Figure 17. Intercomparison of eight models for prediction of mean (τ_m) and maximum (τ_{max}) bed shear-stress due to waves plus a current (reprinted from Soulsby et al. (1993), Coastal Engineering, 21, 41–69, by permission of Elsevier Science Publishers, BV)

Four models which gave a good all-round performance and/or are widely used are the analytical models of Grant and Madsen (1979) and Fredsøe (1984), and the numerical models of Huynh-Thanh and Temperville (1991) and Davies *et al.* (1988). Parametric fitting coefficients for these models are given in Table 9, for use in the method outlined below.

Table 9. Fitting coefficients for wave/current boundary layer models

Model*	a_1	a_2	a_3	a_4	m_1	m_2	m_3	m_4	n_1	n_2	n_3	n_4	I
GM79	0·11	1·95	−0·49	−0·28	0·65	−0·22	0·15	0·06	0·71	−0·19	0·17	−0·15	0·67
F84	−0·06	1·70	−0·29	0·29	0·67	−0·29	0·09	0·42	0·75	−0·27	0·11	−0·02	0·80
HT91	−0·07	1·87	−0·34	−0·12	0·72	−0·33	0·08	0·34	0·78	−0·23	0·12	−0·12	0·82
DSK88	0·05	1·62	−0·38	0·25	1·05	−0·72	−0·08	0·59	0·66	−0·25	0·19	−0·03	0·82

Model*	b_1	b_2	b_3	b_4	p_1	p_2	p_3	p_4	q_1	q_2	q_3	q_4	J
GM79	0·73	0·40	−0·23	−0·24	−0·68	0·13	0·24	−0·07	1·04	−0·56	0·34	−0·27	0·50
F84	0·29	0·55	−0·10	−0·14	−0·77	0·10	0·27	0·14	0·91	0·25	0·50	0·45	3·0
HT91	0·27	0·51	−0·10	−0·24	−0·75	0·13	0·12	0·02	0·89	0·40	0·50	−0·28	2·7
DSK88	0·22	0·73	−0·05	−0·35	−0·86	0·26	0·34	−0·07	−0·89	2·33	2·60	−2·50	2·7
DATA13	0·47	0·69	−0·09	−0·08	−0·53	0·47	0·07	−0·02	2·34	−2·41	0·45	−0·61	8·8
DATA2	1·2	0·0	0·0	0·0	0·0	0·0	0·0	0·0	3·2	0·0	0·0	0·0	0·0

*GM79 = Grant and Madsen (1979); F84 = Fredsøe (1984); HT91 = Huynh-Thanh and Temperville (1991); DSK88 = Davies *et al.* (1988); DATA13, DATA2 = fit to 131 data points using 13,2 coefficients.

A data-based method was also derived by Soulsby (1995a) by optimising the 13 coefficients in the parameterised expression for τ_m which had been used for fitting the theoretical models. The same set of 131 data points was used. This approach gave a significantly better fit to the data than any of the models (although it is a mathematical necessity that it cannot be worse). The coefficients for this method are given as 'DATA13' in Table 9.

It was also found that a two-coefficient optimisation ('DATA2') gave almost as good a fit to the data as the best theoretical models. This method reduces to the simple equation

$$\tau_m = \tau_c \left[1 + 1 \cdot 2 \left(\frac{\tau_w}{\tau_c + \tau_w} \right)^{3 \cdot 2} \right] \tag{69}$$

in which τ_c and τ_w are the bed shear-stresses which would occur due to the current alone and to the wave alone, respectively.

The corresponding expression for τ_{max} (for which there is insufficient data to perform an optimisation) used in the 'DATA13' and the 'DATA2' methods is given by a vector addition of τ_m from Equation (69) and τ_w obtained through Equations (62a) and (57):

$$\tau_{max} = [(\tau_m + \tau_w \cos \phi)^2 + (\tau_w \sin \phi)^2]^{1/2} \tag{70}$$

Calculations of τ_{max} are needed to determine the threshold of motion and entrainment rate of sediments, and of τ_m to determine sediment diffusion.

The methods can be applied equally well to calculating the total bed shear-stress and the skin-friction contribution. In the first case the total z_0 is used, while in the second case a grain-related $z_0 = z_{0s} = d_{50}/12$ is used.

Procedure

1. *Example 5.1. Total shear-stress for waves and currents*

To calculate the maximum (τ_{max}) and mean (τ_m) bed shear-stresses over a wave-cycle, for a sinusoidal wave with:

near-bed orbital velocity amplitude	U_w	0.5 m s^{-1}
and period	T	12.6 s
travelling at angle	ϕ	$30°$
to a steady current of depth-averaged speed	$\bar{U}$	1 m s^{-1}
over a hydrodynamically rough bed with roughness length	z_0	0.001 m
in water of depth	h	10 m
and density	ρ	1027 kg m^{-3}

using the Grant and Madsen (1979) model:

Calculate drag coefficient C_D of steady current in absence of waves by interpolation in Table 10.

$$h/z_0 = 10/0.001 \qquad 10^4$$

$$C_D \qquad 0.00237$$

Calculate wave friction factor f_w for waves in the absence of current by interpolation in Table 10.

$$A/z_0 = 0.5 \times 12.6/(2\pi \times 0.001) \qquad 10^3$$

$$f_w \qquad 0.0316$$

Calculate current-only bed shear-stress $\tau_c = \rho C_D \bar{U}^2$. $\qquad 2.43$ N m^{-2}

Calculate wave-only peak bed shear-stress $\tau_w = \frac{1}{2}\rho f_w U_w^2$. $\qquad 4.06$ N m^{-2}

Calculate $X = \tau_c/(\tau_c + \tau_w)$ $\qquad 0.374$

From Table 9 for the chosen model, calculate coefficients a, m and n, where

$$a = (a_1 + a_2|\cos\phi|^{\mathrm{I}})$$
$$+ (a_3 + a_4|\cos\phi|^{\mathrm{I}})$$
$$\times \log_{10}(f_{\mathrm{w}}/C_{\mathrm{D}})$$

with analogous expressions for m and n:

a	1·04
m	0·680
n	0·575

From Table 9 for the chosen model, calculate coefficients b, p and q, where

$$b = (b_1 + b_2|\cos\phi|^{J})$$
$$+ (b_3 + b_4|\cos\phi|^{J})$$
$$\times \log_{10}(f_{\mathrm{w}}/C_{\mathrm{D}})$$

with analogous expressions for p and q:

b	0·592
p	−0·362
q	0·619

Calculate
$$Z = 1 + aX^m(1 - X)^n \qquad 1·41$$
and
$$Y = X[1 + bX^p(1 - X)^q] \qquad 0·611$$
Calculate
$$\tau_{\max} = Z(\tau_{\mathrm{c}} + \tau_{\mathrm{w}}) \qquad 9·15\ \mathrm{N\ m^{-2}}$$
and
$$\tau_{\mathrm{m}} = Y(\tau_{\mathrm{c}} + \tau_{\mathrm{w}}) \qquad 3·97\ \mathrm{N\ m^{-2}}$$

Table 10. Values of friction factor f_w and drag coefficient C_D from various models

Model*	Values of f_w			
A/z_0	10^2	10^3	10^4	10^5
GM79	0·1057	0·0316	0·0135	0·00690
F84	0·0592	0·0221	0·0102	0·0056
HT91	0·0750	0·0272	0·0121	0·0062
DSK88	0·0701	0·0209	0·0120	0·00661
DATA13, DATA2	0·1268	0·0383	0·0116	0·0035
	Values of C_D			
z_0/h	10^{-2}	10^{-3}	10^{-4}	10^{-5}
GM79, F84,				
DATA13, DATA2	0·01231	0·00458	0·00237	0·00145
HT91	–	0·00482	0·00237	0·00141
DSK88	–	0·00429	0·00222	0·00130

* GM79 = Grant and Madsen (1979); F84 = Fredsøe (1984); HT91 = Huynh-Thanh and Temperville (1991); DSK88 = Davies *et al.* (1988); DATA13, DATA2 from Equation (62a).
[b] Models GM79, F84, DATA2 and DATA13 use the logarithmic profile expression for C_D, Equation (37).

For comparison, the methods of Fredsøe (1984), Huynh-Thanh and Temperville (1991) and DATA13 give, respectively, $\tau_{max} = 7·94$, $8·27$, $7·90 \, \mathrm{N\,m^{-2}}$ and $\tau_m = 2·91$, $3·04$, $3·26 \, \mathrm{N\,m^{-2}}$. The values obtained from the DATA2 method (Equations (70) and (69)) are $\tau_{max} = 7·89 \, \mathrm{N\,m^{-2}}$ and $\tau_m = 3·24 \, \mathrm{N\,m^{-2}}$.

2. The methods of Huynh-Thanh and Temperville (1991), Grant and Madsen (1979), Fredsøe (1984) and 'DATA13' (Soulsby 13 in the menu) are most easily calculated using SandCalc under Hydrodynamics–Waves & Currents–Total shear-stress. Sand-Calc uses a smooth interpolation procedure based on Table 10 to calculate C_D and f_w for the Wave & Currents–Total shear-stress methods.

Threshold of motion

6

6. *Threshold of motion*

6.1.   GENERAL

The threshold of motion of sediments on the seabed is an important factor in most types of computation concerned with sediment response to currents and/or waves. It is particularly required in applications involving: scour (and scour protection measures) around structures; sea bed mobility calculations connected with licensing of aggregate extraction; bedload transport (especially of coarser sediments); and entrainment of finer sediments into suspension.

6.2.   THRESHOLD CURRENT SPEED

Knowledge

For very slow flows over a sand bed the sand remains immobile. If the flow velocity is slowly increased, a velocity is reached at which a few grains begin to move. This is called the *threshold (or initiation) of motion* or *incipient motion*. A similar process occurs beneath waves, and beneath combined waves and a current.

For a steady current, the threshold (or critical) depth-averaged speed $\bar{U}_{cr}$ required to move a grain of diameter d on a flat, horizontal, un-rippled bed in water of depth h can be predicted by a number of methods.

Van Rijn (1984) gives the following formula, valid for fresh water at $15\,^{\circ}\mathrm{C}$, $\rho_s = 2650\,\mathrm{kg\,m^{-3}}$ and $\mathrm{g} = 9.81\,\mathrm{m\,s^{-2}}$:

$$\bar{U}_{cr} = 0.19(d_{50})^{0.1}\log_{10}(4h/d_{90})$$

$$\text{for } 100 \leq d_{50} \leq 500\,\mu\mathrm{m} \qquad \text{SC } (71a)$$

$$\bar{U}_{cr} = 8 \cdot 5(d_{50})^{0 \cdot 6} \log_{10}(4h/d_{90})$$

$$\text{for } 500 \le d_{50} \le 2000 \, \mu\text{m} \qquad \text{SC (71b)}$$

in which all units must be in metres and seconds, and d_{50} and d_{90} are as defined in Section 2.2.

The expression for threshold bed shear-stress (see Section 6.4) given by Equation (77) can be combined with the friction law given by Equation (34) to give the Soulsby formula for threshold current speed, valid for any non-cohesive sediment and water conditions for which $D_* > 0 \cdot 1$, and valid in any units:

$$\bar{U}_{cr} = 7 \left(\frac{h}{d_{50}} \right)^{1/7} [g(s-1)d_{50}f(D_*)]^{1/2},$$

$$\text{for } D_* > 0 \cdot 1 \qquad \text{SC (72a)}$$

with

$$f(D_*) = \frac{0 \cdot 30}{1 + 1 \cdot 2 D_*}$$

$$+ \, 0 \cdot 055[1 - \exp(-0 \cdot 020 D_*)] \qquad \text{SC (72b)}$$

$$D_* = \left[\frac{g(s-1)}{\nu^2} \right]^{1/3} d_{50}$$

$$s = \text{ratio of densities of grain and water}$$

$$\nu = \text{kinematic viscosity of water.}$$

Curves showing $\bar{U}_{cr}$ as a function of grain diameter given by Equation (72a,b) are shown in Figure 18 for a range of water depths, for the specific case of quartz grains in sea water at $10\,°C$ and 35 ppt. The same curves can be used to obtain the threshold velocity $U_{cr}(z)$ at a specific height z by using the relationship $\bar{U}_{cr} = U_{cr}(z = 0 \cdot 32h)$ as given by Equation (28a), where $\bar{U}_{cr}$ is given by Equation (72). These values of $U_{cr}(z)$ are also marked in Figure 18.

Procedure

1. The threshold depth-averaged (steady) current speed $\bar{U}_{cr}$ can be obtained from Equation (71) or Equation (72). Inputs are

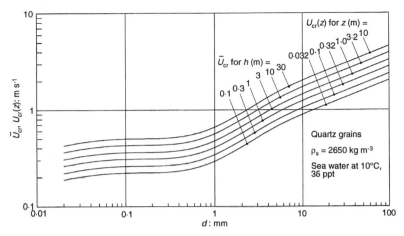

Figure 18.   Threshold current speed for motion of sediment by steady flows

median grain diameter d_{50}, 90th percentile grain diameter d_{90} (for Equation (71)) and water depth h. For Equation (72), the water density ρ, viscosity ν and sediment density ρ_s are also needed.

Example 6.1.   Threshold current speed

Calculate the threshold depth-averaged current speed for steady flow over a flat sand bed with the following characteristics:

$$d_{50} = 200 \text{ μm}, \ d_{90} = 300 \text{ μm}, \ h = 5 \text{ m},$$

$$\rho = 1027 \text{ kg m}^{-3}, \ \nu = 1\cdot36 \times 10^{-6} \text{ m}^2\text{ s}^{-1},$$

$$\rho_s = 2650 \text{ kg m}^{-3}.$$

From Equation (71a), $\bar{U}_{cr} = 0\cdot39 \text{ m s}^{-1}$.
From Equation (72), $\bar{U}_{cr} = 0\cdot39 \text{ m s}^{-1}$ (also).
Larger differences between the methods are found for other grain sizes.

Although Equation (71) is simpler, Equation (72) is recommended because: it covers a wider range of grain sizes; it allows for variation in ρ, ν and ρ_s; and it is completely consistent with the threshold criterion for waves given by Figure 19.

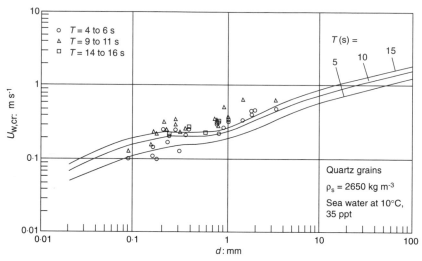

Figure 19. Threshold orbital velocity for motion of sediment by waves

6.3. THRESHOLD WAVE HEIGHT

Knowledge

Under waves, the threshold of motion of sand depends on the bottom orbital velocity amplitude U_w (see Section 4.4), the wave period T, and the grain diameter d and density ρ_s. The threshold orbital velocity U_{wcr} can be determined from the equations of Komar and Miller (1974):

$$U_{wcr} = [0{\cdot}118g(s-1)]^{2/3}d^{1/3}T^{1/3} \quad \text{for } d < 0{\cdot}5 \text{ mm}$$

SC (73a)

$$U_{wcr} = [1{\cdot}09g(s-1)]^{4/7}d^{3/7}T^{1/7} \quad \text{for } d > 0{\cdot}5 \text{ mm}$$

SC (73b)

where g = acceleration due to gravity
s = ratio of densities of grain and water

These equations are widely used, but have the disadvantage of having a large discontinuity at $d = 0{\cdot}5$ mm.

The threshold orbital velocity can also be deduced from Equation (77) for threshold bed shear-stress (see Section 6.4). The method of Soulsby is given by the curves shown in Figure 19, which give U_{wcr} as a function of grain diameter for a range of wave periods, for the specific case of quartz grains in sea water at 10 °C and 35 ppt. They were computed at each grain size by increasing the wave orbital velocity U_w in small steps, and converting to bed shear-stress by the procedure given in Section 4.5. When the computed bed shear-stress is equal to the threshold value from Equation (77), then $U_{wcr} = U_w$. It is not possible to write an analytical formula for this, because of the complexity of including laminar, smooth turbulent and rough turbulent expressions for the wave friction factor. Experimental data taken from the compilation by Van Rijn (1989) are shown in Figure 19. The theoretical curves in some cases lie below the data points, possibly because an average rather than a peak stress due to the waves may be more representative for comparisons with thresholds under currents. There is an increase in U_{wcr} with wave period indicated by the theoretical curves (Equations 73a,b show a similar trend with T), although the data only weakly support this.

Procedure

1. The threshold orbital velocity U_{wcr} under waves of period T for a grain of diameter d can be obtained from Figure 19. The orbital velocity can be converted to wave height using the methods described in Section 4.4.

Example 6.2. Threshold orbital velocity

Calculate the threshold orbital velocity for monochromatic waves passing over a flat sand bed with the following characteristics:

$d = 200$ μm, $T = 8$ s.

From Figure 19, the threshold predicted by the method of Soulsby is $U_{wcr} = 0.17$ m s^{-1}.

Using Equation (73a), the threshold predicted by the method of Komar and Miller (1974) is $U_{wcr} = 0.18$ m s^{-1}.

Larger differences between the methods are found for coarser grain sizes.

2. The method of Soulsby (Figure 19) is recommended, because: it is continuous throughout the range of grain sizes; it covers smooth turbulent, as well as laminar and rough turbulent, conditions; and it is fully compatible with the threshold for currents given by Equation (72).

3. The methods of Komar and Miller (1974) and Soulsby are available in SandCalc under Sediments–Threshold–Waves.

6.4. THRESHOLD BED SHEAR-STRESS

Knowledge

A more precise measure of the threshold of motion can be given in terms of the bed shear-stress (see Section 3.3). This approach was developed by Shields (1936) in terms of the ratio of the force exerted by the bed shear-stress acting to move a grain on the bed, to the submerged weight of the grain counteracting this. The threshold Shields parameter θ_{cr}, is defined as

$$\theta_{cr} = \frac{\tau_{cr}}{g(\rho_s - \rho)d} \tag{74}$$

where τ_{cr} = threshold bed shear-stress
 g = acceleration due to gravity = $9\cdot81\,\mathrm{m\,s^{-1}}$
 ρ_s = grain density
 ρ = water density
 d = grain diameter

It can be plotted against the dimensionless grain size D_* given by

$$D_* = \left[\frac{g(s-1)}{\nu^2}\right]^{1/3} d \tag{75}$$

where ν = kinematic viscosity of water

 $s = \rho_s/\rho$

Shields originally plotted the data available to him (all for currents) in the form θ_{cr} versus the grain Reynolds number $u_{*cr}\,d/\nu$, where $u_{*cr} = (\tau_{cr}/\rho)^{1/2}$. However, this is inconvenient to use, because the unknown u_{*cr} appears on both axes. A direct mathematical transformation can be made to a plot of θ_{cr} versus D_*, which is easier to use in practical applications.

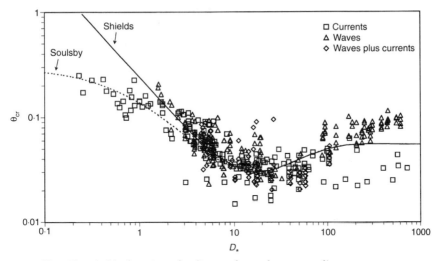

Figure 20. Threshold of motion of sediments beneath waves and/or currents

The classical work of Shields for steady currents can be extended to waves, and combined waves and currents, as shown in Figure 20. The values of θ_{cr} for wave data are plotted as the amplitude θ_w of the oscillatory Shields parameter, and for waves-plus-currents are plotted as the maximum value θ_{max} over a wave-cycle.

Data for natural sand in water and data for other grains/fluids collapse together with this non-dimensionalisation, and the results for currents, waves and combined waves and currents also show relatively similar behaviour. It is seen that for large grain sizes (gravel) the experimental values of θ_{cr} for waves are significantly larger than for currents. The reason for this is not clear, but may partly be because an average rather than the peak stress would be more appropriate.

Shields plotted a curve by hand through the limited amount of data on threshold under currents available to him in the 1930s. An algebraic expression that fits Shields' curve closely is proposed by Soulsby and Whitehouse (1997):

$$\theta_{cr} = \frac{0.24}{D_*} + 0.055[1 - \exp(-0.020D_*)] \qquad \text{SC (76)}$$

It can be seen (Figure 20) that this curve passes reasonably well through waves and wave-plus-current data, as well as the extensive set of currents data now available. However, for very fine grain sizes, Equation (76) greatly overpredicts the data. Force considerations by Bagnold showed that θ_{cr} cannot exceed a value of about 0·30, because this exerts a sufficient force on the grains to overcome the weight of every grain in the topmost layer of the bed. A correction to account for this has been made to Equation (76) by Soulsby and Whitehouse (1997) to give an improved threshold bed shear-stress formula (see Figure 20):

$$\theta_{cr} = \frac{0\cdot30}{1 + 1\cdot2D_*} + 0\cdot055[1 - \exp(-0\cdot020D_*)] \qquad \text{SC (77)}$$

Figure 21 shows curves of τ_{cr} versus d for the case $g = 9\cdot81 \text{ m s}^{-1}$, $\rho_s = 2650 \text{ kg m}^{-3}$, temperature $= 10\,^{\circ}\text{C}$, salinity $= 35$ ppt, typical of sand in sea water, and for the case of sand in fresh water at 20 °C.

It must be noted that Figures 18–21 and Equations (76) and (77) apply to the threshold of motion on an initially flat horizontal bed. If the bed is rippled, these results apply only to the skin friction component of the bed shear-stress τ_{0s} (see Section 1.4).

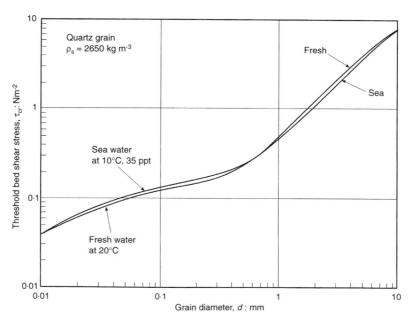

Figure 21. Threshold bed shear-stress for motion of quartz grains of sieve diameter d

Equations (76) and (77) both give an approximately constant value of $\theta_{cr} = 0.055$ for large grain sizes ($D_* > 200$). For these larger grain sizes, corresponding to $d > 10\,mm$ for quartz grains in sea water, a formula can be derived for the threshold grain diameter d_{cr}, which is just immobile for given flow conditions. This is useful, for example, for calculating the size of rock or gravel scour protection material.

A formula for d_{cr} for *steady flow* can be obtained by combining $\theta_{cr} = 0.055$ with Equations (74) and (34), to give

$$d_{cr} = \frac{0.250\bar{U}^{2.8}}{h^{0.4}[g(s-1)]^{1.4}} \qquad \text{for } d_{cr} > 10\,mm \qquad (78)$$

Similarly, a formula for d_{cr} for *waves* can be obtained by combining $\theta_{cr} = 0.055$ with Equations (74) and (62a), to give

$$d_{cr} = \frac{97.9U_w^{3.08}}{T^{1.08}[g(s-1)]^{2.08}} \qquad \text{for } d_{cr} > 10\,mm \qquad (79)$$

where d_{cr} = grain diameter which is just immobile for a given flow
$\bar{U}$ = depth-averaged current speed
h = water depth
U_w = water orbital velocity amplitude at seabed
T = period of water wave
g = acceleration due to gravity
s = ratio of densities of grain and water

If the bed is sloping, then gravity provides a component of force on the grain which may increase or decrease the threshold shear-stress required from the flow. The gravity force can be added vectorially to the shear-stress force from the flow to calculate the threshold condition for a grain on a bed of arbitrary streamwise and cross-stream slopes. The threshold bed shear-stress $\tau_{\beta cr}$ for sand grains on a bed sloping at angle β to the horizontal, in a flow making an angle ψ to the upslope direction (see Figure 22c) is related to the value τ_{cr} for the same grains on a horizontal bed by the expression:

$$\frac{\tau_{\beta cr}}{\tau_{cr}} = \frac{\cos\psi\sin\beta + (\cos^2\beta\tan^2\phi_i - \sin^2\psi\sin^2\beta)^{1/2}}{\tan\phi_i}$$

SC (80a)

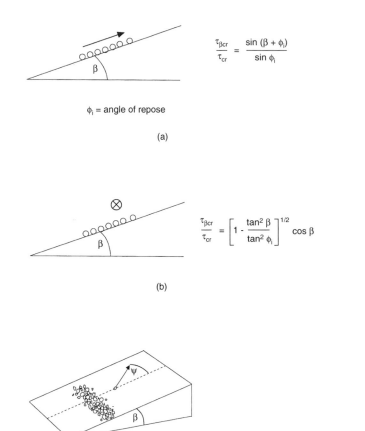

Figure 22. *Threshold of motion on sloping beds: (a) longitudinal slope, β;
(b) lateral slope, β; (c) general slope — flow at angle ψ to slope of angle β*

The angle ϕ_i is the angle of repose of the sediment (see Section
2.2) at which slope the sediment will avalanche in a zero flow.
Thus, if $\beta > \phi_i$, then avalanching occurs. If the flow is up the
slope ($\psi = 0°$, see Figure 22a), then Equation (80a) reduces to:

$$\frac{\tau_{\beta cr}}{\tau_{cr}} = \frac{\sin(\phi_i + \beta)}{\sin \phi_i} \qquad (80b)$$

If the flow is down the slope ($\psi = 180°$), then

$$\frac{\tau_{\beta cr}}{\tau_{cr}} = \frac{\sin(\phi_i - \beta)}{\sin \phi_i} \qquad (80c)$$

If the flow is directed laterally across the slope ($\psi = \pm 90°$, see Figure 22b), then

$$\frac{\tau_{\beta cr}}{\tau_{cr}} = \cos \beta \left(1 - \frac{\tan^2 \beta}{\tan^2 \phi_i}\right)^{1/2} \qquad (80d)$$

Procedure

1. The threshold bed shear-stress τ_{cr} of a well-sorted sediment (see Section 2.2) can be calculated from Figure 21, *or*, more accurately, from Equation (77).

Example 6.3. Threshold bed shear-stress

Obtain temperature in degrees C		10
Obtain salinity in ppt		35
Calculate kinematic viscosity in m^2 s^{-1} (Example 2.1)	ν	$1 \cdot 36 \times 10^{-6}$
Obtain grain density in kg m^{-3}	ρ_s	2650
Calculate water density in kg m^{-3} (Example 2.1)	ρ	1027
Calculate $s = \rho_s/\rho$		2·58
Obtain grain diameter in m	d	200×10^{-6}
Calculate D_* from Equation (75)		4·06
Calculate threshold Shields parameter by Soulsby's method from Equation (77)	θ_{cr}	0·0553

Calculate threshold bed
shear-stress from
Equation (74)

$$\tau_{cr} = 0.0553 \times 9.81$$
$$\times (2650 - 1027) \times 200 \times 10^{-6} \qquad = 0.176 \, \mathrm{N\,m^{-2}}$$

For comparison, the Shields curve (Equation (76)) gives $\theta_{cr} = 0.0633$ and $\tau_{cr} = 0.201 \, \mathrm{N\,m^{-2}}$.

The two methods give similar results (within about 10%) for grains coarser than about 200 µm, but for finer grains Equation (77) fits the data more closely.

2. If the bed is sloping, follow one of the methods given above to obtain τ_{cr}, and then use Equation (80a) to obtain $\tau_{\beta cr}$.

Example 6.4. Threshold on sloping bed

Take the same input values as given in Example 6.3, which yielded $\tau_{cr} = 0.176 \, \mathrm{N\,m^{-2}}$, but now calculate the threshold bed shear-stress for a current flowing at an angle of 45° to the upslope direction on a bed sloping at 20°. Take the angle of repose $\phi_i = 32°$. Then Equation (80a) with $\beta = 20°$, $\psi = 45°$, $\phi_i = 32°$ yields $\tau_{\beta cr}/\tau_{cr} = 1.24$, so that $\tau_{\beta cr} = 1.24 \times 0.176 = 0.219 \, \mathrm{N\,m^{-2}}$.

3. To calculate the grain diameter d_{cr}, which is just immobile for a given flow, Figure 18 may be used for currents, and Figure 19 for waves. For grains larger than 10 mm, Equation (78) can be used for currents and Equation (79) for waves.

If the bed shear-stress τ is known, then Figure 21 may be used to obtain d_{cr}.

Example 6.5. Critical grain size

In a combined wave and current flow, the peak bed shear-stress has been calculated to be $2.0 \, \mathrm{N\,m^{-2}}$. What size of sand or gravel will be stable in this flow, assuming quartz grains in sea water at 10 °C and 35 ppt? From Figure 21, for $\tau_{cr} = 2.0 \, \mathrm{N\,m^{-2}}$, the threshold grain diameter is 3.0 mm.

Bed features

7

7. Bed features

7.1. GENERAL

A universal characteristic of flows in rivers, estuaries and the sea is the tendency of a sandy bed to form itself into one of a variety of types of bed feature. The type of feature (or bedform) depends on the strength and nature of the flow: steady current, tidal current, waves, or a combination of these. Some of the types of bedforms are illustrated in Figure 23. Steady currents in rivers form small ripples, large dunes, and sometimes ripples on the back of dunes. Similar features are formed by tidal currents in estuaries and the sea, where the oscillatory nature of the tides continually modifies the ripple pattern, and may cause a more symmetrical form of dune or sandwave.

The nomenclature of the larger features in the sea is not well established. The name sandwave is used by many oceanographers for all the large transverse features including ones similar to those called dunes in rivers, and this nomenclature is followed in this book. However, hydraulic engineers often reserve the name sandwaves for the very largest features (hundreds or thousands of metres wavelength) found in the sea. The names sandwaves and bedwaves are also sometimes used to mean any types of wavy disturbance of the bed, including ripples.

A wide variety of other features, both longitudinal and transverse to the flow, are also found in the sea. Where waves are the dominant hydrodynamic forcing agent, wave-generated ripples are formed with a distinctively different shape to current-generated ripples. Waves can also give rise to larger bed features, such as breaker bars in the surf zone (Figure 23d).

Applications involving bed features include smothering of water intakes, and undermining of pipelines leading to 'spanning' and possible breakage. Bed features also have a dominant influence on the frictional characteristics and turbulence

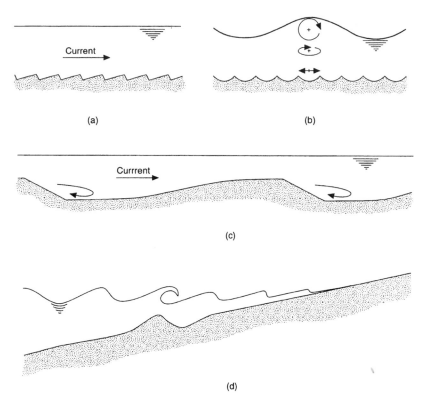

Figure 23. Types of bedforms: (a) current ripples; (b) wave ripples; (c) sandwaves/dunes; (d) breaker bar

formation in flows, and have both a direct (bedform migration) and indirect (enhanced suspension) effect on sediment transport.

7.2.   CURRENT RIPPLES AND SANDWAVES

Knowledge

For flows which exceed the threshold of motion, an initially flat bed may deform into various types of bed feature, ranging in size from small ripples up to major sandbanks. At high flow speeds, and where sand supply is limited, the sand may be confined to *crescentic dunes* and/or streamwise *sand ribbons* moving over a gravel pavement. At more moderate velocities the bedforms are orientated transverse to the flow, and may comprise *ripples*, *dunes* and/or *sandwaves*. *Sandbanks* form in response to large-scale current patterns.

Ripples are small bed features whose height and wavelength are small compared to the water depth. They form on sandy beds with grain sizes up to about 0·8 mm, for flow speeds which are above the threshold of motion but not so intense that the ripples are washed out.

Grains coarser than about 0·8 mm do not form ripples, although sandwaves can form in all grain sizes, including gravel, when they are called gravel waves. At very high flow speeds (for example, $\bar{U} > 1·5 \, \text{m s}^{-1}$ for $d = 0·2 \, \text{mm}$) the ripples are washed out to leave a flat bed with a sheet flow of intense sediment transport. At very low flow speeds, below the threshold of motion, the bed features will retain the form they had at the time when the flow fell below the threshold value. This form is generally not known when calculations are made at a single point in time without considering the previous flow history, so for definiteness it is often assumed that the bed is flat for conditions below threshold. However, in reality it is more likely that a ripple pattern will remain frozen if the flow speed decreases reasonably slowly, so a better guess for calculating the bed roughness for flows below threshold is to treat it as rippled. In areas of intense biological activity, on the other hand, a rippled bed can be levelled by burrowing animals in only a few hours.

Current-generated ripples are asymmetrical in cross-section with the steeper slope on the downstream side of the crest (Figure 23a). They form an irregular, strongly three-dimensional, pattern when viewed from above, with the length of the crest of an individual ripple being typically only one to three times the wavelength. Their wavelength, λ_r, is often quoted as being approximately 1000 grain diameters, and their height, Δ_r, up to about 1/7 of the wavelength:

$$\lambda_r = 1000 d_{50} \tag{81a}$$

$$\Delta_r = \lambda_r/7 \tag{81b}$$

An average estimate for all grain sizes based on observations at the sea bed, on inter-tidal sand flats, and in laboratory flumes, is that current generated ripples have a wavelength of about 0·14 m and a height of about 0·016 m. The ripples move downstream slowly in the direction of the current. In a strong tidal flow (spring tides) the ripple asymmetry, orientation and migration change with the changing current direction.

Dunes and *sandwaves* are much larger current-generated features, often tens of metres in wavelength, and some metres in height (Figure 23c). Their heights and wavelengths are governed by the water depth as well as the bed shear-stress. A rough guide to their wavelength is 6 × (water depth). There are various empirical formulae for the height Δ_s and wavelength λ_s of sandwaves, of which the most reliable are those of:

Yalin (1964):

$$\Delta_s = 0 \quad \text{for } \tau_{0s} < \tau_{cr} \qquad\qquad \text{SC}(82a)$$

$$\Delta_s = \frac{h}{6}\left(1 - \frac{\tau_{cr}}{\tau_{0s}}\right) \quad \text{for } \tau_{cr} \leq \tau_{0s} < 17\cdot6\tau_{cr} \qquad \text{SC}(82b)$$

$$\Delta_s = 0 \quad \text{for } \tau_{0s} \geq 17\cdot6\tau_{cr} \qquad\qquad \text{SC}(82c)$$

$$\lambda_s = 2\pi h \qquad\qquad\qquad \text{SC}(82d)$$

Van Rijn (1984)

$$\Delta_s = 0 \quad \text{for } \tau_{0s} < \tau_{cr} \qquad\qquad \text{SC}(83a)$$

$$\Delta_s = 0\cdot11h\left(\frac{d_{50}}{h}\right)^{0\cdot3}(1 - e^{-0\cdot5T_s})(25 - T_s)$$

$$\qquad\qquad\qquad \text{for } \tau_{cr} < \tau_{0s} < 26\tau_{cr} \quad \text{SC}(83b)$$

$$\Delta_s = 0 \quad \text{for } \tau_{0s} > 26\tau_{cr} \qquad\qquad \text{SC}(83c)$$

$$\lambda_s = 7\cdot3h \qquad\qquad\qquad \text{SC}(83d)$$

where Δ_s = height of sandwave
$\quad\;\; \lambda_s$ = wavelength of sandwave
$\quad\;\; h$ = water depth
$\quad\;\; \tau_{0s}$ = bed shear-stress due to skin friction
$\quad\;\; \tau_{cr}$ = threshold bed shear-stress for sediment motion
$\quad\;\; T_s = \dfrac{\tau_{0s} - \tau_{cr}}{\tau_{cr}}$
$\quad\;\; d_{50}$ = median grain size

These formulae are illustrated in Figure 24. The formula of Van Rijn is recommended, as it has been calibrated against the largest data set. Note that Van Rijn uses $k_s = 3d_{90}$ to calculate τ_{0s}, yielding a rather larger value than that obtained with $k_s = 2 \cdot 5d_{50}$.

Equations (82) and (83) apply to steady unidirectional flows, as found in rivers. In tidal conditions, where the current speed is always changing, the bedforms cannot fully adapt to the flow, and the formulae are less reliable. In a unidirectional flow the bedforms migrate slowly downstream. A similar behaviour is observed in estuaries and the sea if there is strongly dominant flow in the flood or the ebb direction. An example is shown in Figure 25, in which sandwaves of length 15 m and height 0·8 m migrate up to 1 m per day in a part of the Taw Estuary, south-west England, with a strongly dominant flood tide.

The term dunes tends to be used for the large bedforms in rivers, while the term sandwaves is used in the sea. It is not entirely clear whether they are morphologically identical, but in the sea sandwaves (which can be hundreds of metres in length) often have bedforms of an intermediate size superimposed on them, and these are then referred to as dunes. In addition, ripples can co-exist with dunes and/or sandwaves.

Bedform migration can be used as a method of measuring the bedload transport rate. If it is assumed that all mobile grains roll over the bedform, up the upstream (stoss) face and down the downstream (lee) face, and come to rest in the trough, then the

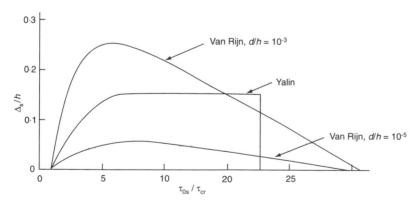

Figure 24.   Formulae for sandwave height

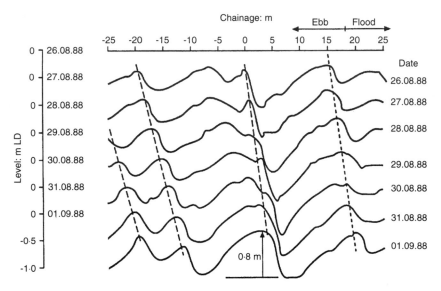

Figure 25. Sandwave migration in Taw Estuary — daily bed surveys; vertical exaggeration = 10 ×

volumetric bedload transport rate q_b can be calculated from the equation

$$q_b = a_m \Delta V_{mig} \tag{84}$$

where a_m = constant
Δ = height of bedform
V_{mig} = migration speed

The constant a_m is the product of $(1 - \varepsilon)$, where ε is porosity (see Section 2.3), and a factor describing the shape of the bedform. If $\varepsilon = 0.40$ and the shape is triangular, then $a_m = 0.60 \times 0.5 = 0.30$. Observed values are generally in the range $0.22 < a_m < 0.37$. Use a value of $a_m = 0.32$ if the shape and porosity are not known (Jinchi, 1992).

The above method can be used either for ripples or for dunes/sandwaves.

The assumption on which Equation (84) is based may not be entirely valid, since many grains do not come to rest in the trough, but either continue to roll along the bed, or are carried into suspension. Hence measurements of bedload transport rate from bedform migration may be underestimates by up to about a factor of two.

At high current velocities the ripples and dunes are washed out, and the bed becomes flat with intense sediment transport taking place as a slurry or *sheet flow* in a thickness of a few millimetres above the bed. This condition takes place according to the approximate criterion

$$\theta_s > 0{\cdot}8 \tag{85a}$$

or

$$\tau_{0s} > 0{\cdot}8g\rho(s-1)d \tag{85b}$$

where τ_{0s} = skin-friction bed shear-stress
$\quad\ \theta_s$ = skin-friction Shields parameter
$\quad\ g$ = acceleration due to gravity
$\quad\ \rho$ = water density
$\quad\ s$ = relative density of sediment
$\quad\ d$ = grain diameter

In the sea, ripple wash-out occurs in shallow water with strong currents, or under intense wave action such as in the surf zone.

Procedure

1. *Example 7.1. Sandwave dimensions*

   To calculate sandwave dimensions given the flow conditions, obtain values of:

water depth	h	10 m
grain diameter	d	200 µm
maximum tidal current velocity	$\bar{U}$	$1{\cdot}0$ m s^{-1}
Calculate threshold bed shear-stress (Example 6.3)	τ_{cr}	$0{\cdot}176$ N m^{-2}
Calculate actual bed shear-stress due to skin-friction from Equation (34)	τ_{0s}	$0{\cdot}952$ N m^{-2}

Use Van Rijn method

Calculate $T_s = (\tau_{0s} - \tau_{cr})/\tau_{cr}$ 4·41

Since $\tau_{cr} \leq \tau_{0s} \leq 26\tau_{cr}$, use
Equation (83b) to calculate Δ_s 0·78 m

The wavelength given by
   Equation (83d) is λ_s 73 m

    Yalin's formula (Equation (82)) gives an alternative answer of $\Delta_s = 1·36$ m and $\lambda_s = 63$ m.

2. To measure bedload transport from sandwave migration, make repeated echo soundings along the dominant current direction, or, in the inter-tidal zone, make repeated staff-and-level surveys along a line perpendicular to the sandwave crests. In both cases, great accuracy in position fixing is required. For semidiurnal tides a period of 12·5 or 25 h between surveys is suitable.

Example 7.2. Sandwave migration

Analyse the records to give:

mean trough-to-crest height Δ 0·8 m

mean migration speed, by
overlaying successive profiles
and sliding to obtain best fit V_{mig} 1·0 m/day

Use Equation (84) with
$a_m = 0·32$ to obtain
volumetric transport
rate $q_b = 0·32 \times 0·8 \times 1·0$ $= 0·26$ m^2/day
Mean transport rate
averaged over a day $= 0·26/(24 \times 3600)$
     $= 3·0 \times 10^{-6}$ m^2 s^{-1}

7.3.   WAVE RIPPLES

Knowledge

Wave-generated ripples are symmetrical about the crest in cross-section, with the crest being relatively sharp (Figure 23b). Their crests are aligned with the crests of the water waves, and, when viewed from above, form a regular pattern of almost parallel lines with very long crest lengths, occasionally broken up by bifurcations. Their wavelength λ_r is typically between 1·0 and 2·0 times the orbital amplitude, $A = U_w T/(2\pi)$, of the wave motion at the bed, where U_w is the orbital velocity amplitude and T is the wave period. Their height Δ_r is typically between 0·1 and 0·2 times their wavelength.

Wave ripples are washed out by very large orbital velocities, to leave a flat bed with oscillatory sheet flow. The criterion for ripple wash-out is given in terms of either the skin-friction Shields parameter, θ_{ws}, with a typical critical value of about 0·8 (cf. Equation (85a)), or the wave mobility number Ψ, with a typical critical value of about 150, where

$$\theta_{ws} = \frac{\tau_{ws}}{g(\rho_s - \rho)d} \tag{86}$$

$$\Psi = \frac{U_w^2}{g(s-1)d} \tag{87}$$

Various methods have been proposed to calculate Δ_r and λ_r. Grant and Madsen (1982):
For $\theta_{ws} \leq \theta_{cr}$,

$$\Delta_r = \lambda_r = 0 \text{ (or pre-existing values)} \tag{SC (88a)}$$

For $\theta_{cr} < \theta_{ws} \leq \theta_B$,

$$\Delta_r = 0\cdot22(\theta_{ws}/\theta_{cr})^{-0\cdot16}A \tag{SC (88b)}$$

$$\lambda_r = \Delta_r/[0\cdot16(\theta_{ws}/\theta_{cr})^{-0\cdot04}] \tag{SC (88c)}$$

For $\theta_{ws} > \theta_B$,

$$\Delta_r = 0\cdot48(D_*^{1\cdot5}/4)^{0\cdot8}(\theta_{ws}/\theta_{cr})^{-1\cdot5}A \tag{SC (88d)}$$

$$\lambda_r = \Delta_r/[0\cdot28(D_*^{1\cdot5}/4)^{0\cdot6}(\theta_{ws}/\theta_{cr})^{-1\cdot0}] \tag{SC (88e)}$$

with

$$\theta_B = 1\cdot8\theta_{cr}(D_*^{1\cdot5}/4)^{0\cdot6} \tag{SC (88f)}$$

$$D_* = \left[\frac{g(s-1)}{\nu^2}\right]^{1/3}d$$

Nielsen (1992) gave for laboratory (regular) waves

For $\theta_{ws} \leq \theta_{cr}, \Delta_r = \lambda_r = 0$ (or pre-existing values)

$$\text{SC (89a)}$$

$$\Delta_r = (0.275 - 0.022\Psi^{0.5})A \quad \text{for } \Psi < 156 \qquad \text{SC (89b)}$$

$$\lambda_r = \Delta_r/(0.182 - 0.24\theta_{ws}^{1.5}) \quad \text{for } \theta_{ws} < 0.831 \qquad \text{SC (89c)}$$

For $\Psi \geq 156$ or $\theta_{ws} \geq 0.831, \Delta_r = \lambda_r = 0 \qquad \text{SC (89d)}$

The wash-out conditions $\Psi = 156$ and $\theta_{ws} = 0.831$ are not entirely compatible with each other.

A more detailed method, based on a mass of data, was proposed by Mogridge et al. (1994).

Procedure

1. To calculate the height Δ_r and wavelength λ_r of wave-ripples on a quartz sand bed in sea water at $10\,°C$ and $35\,ppt$, first obtain the height H and period T of the waves. The waves are assumed to be monochromatic.

Example 7.3. Wave ripples

Obtain wave height	H	1 m
Obtain wave period	T	6 s
Obtain water depth	h	10 m
Calculate orbital velocity, using Figure 14 (monochromatic)	U_w	0.310 m s^{-1}
Calculate orbital amplitude $0.310 \times 6/2\pi$	A	0.296 m
Obtain grain size of bed	d_{50}	0.2 mm
Calculate threshold Shields parameter (see Example 6.3) from Shields curve	θ_{cr}	0.0633
Calculate friction factor using Swart Equation (60)	f_{wr}	0.0118
Calculate skin-friction Shields parameter	θ_{ws}	0.183

Calculate wave mobility parameter (Equation (87))	Ψ	31·0
Calculate ripple height using Nielsen method (Equation (89b))	Δ_r	0·0452 m
Calculate ripple wavelength using (Equation (89c))	λ_r	0·265 m

2. For comparison, the Grant and Madsen method gives $\Delta_r = 0·0579\,\text{m}$, $\lambda_r = 0·373\,\text{m}$.

7.4. FRICTION DUE TO BEDFORMS

Knowledge

When ripples, dunes or sandwaves are present, they generate form drag due to the pattern of dynamic pressure distribution over their surface. This is bluff-body drag, analogous to the wind resistance of a car. The form drag may be many times larger than the skin friction acting on the sand grains, and is often the dominant cause of resistance felt by a river, or tidal currents in estuaries and the sea.

For sediment transport purposes, the skin friction τ_{0s} is responsible for bedload transport and entrainment of sand from the bed, while the form drag τ_{0f} is associated with intense turbulence which diffuses the suspended sediment up into the flow. Further discussion of τ_{0s} and τ_{0f}, and their addition to give the total stress τ_0, is given in Section 1.4, Section 3.4 and Equation (39).

For *current-generated ripples*, the total stress is usually obtained by means of the roughness length z_0, or the Nikuradse roughness k_s, where $k_s = 30z_0$. Table 7 gives a mean value from measurements over rippled sand beds in the sea of $z_0 = 6\,\text{mm}$. Alternatively, the form-drag component, z_{0f}, can be related to the height Δ_r and wavelength λ_r of the ripples:

$$z_{0f} = a_r \frac{\Delta_r^2}{\lambda_r} \qquad\qquad \text{SC (90)}$$

where various investigations give a_r in the range $0.3 < a_r < 3$, with a typical value of $a_r = 1.0$.

The total roughness length z_0 is then obtained using Equation (43), where a sediment-transport component, z_{0t}, may also be appropriate.

Friction over *wave-generated ripples* can be tackled in a similar way. Equation (90) is used with an appropriate value of a_r to give z_{0f}, and a sediment-transport component may also be added. Several methods have been proposed.

Grant and Madsen (1982) used $a_r = 0.923$ in Equation (90), with Δ_r and λ_r calculated by their own method (Equation (88)), plus a sediment-transport component of

$$z_{0t} = 5.33(s + 0.5)d_{50}\theta_{cr}\left[\left(\frac{\theta_{ws}}{\theta_{cr}}\right)^{0.5} - 0.7\right]^2 \qquad \text{SC (91)}$$

Nielsen (1992) used $a_r = 0.267$ in Equation (90), with Δ_r and λ_r calculated by his own method (Equation (89)), plus a sediment-transport component of

$$z_{0t} = 5.67(\theta_{ws} - 0.05)^{0.5}d_{50} \qquad \text{SC (92)}$$

Raudkivi (1988) used $a_r = 0.533$ in Equation (90), with Δ_r and λ_r calculated by Nielsen's method (Equation (89)), plus a sediment-transport component of

$$z_{0t} = 0.00533U_w^{2.25} \qquad \text{SC (93)}$$

where z_{0t} is in metres and U_w in m s^{-1}.

Friction due to *dunes* and *sandwaves* can be treated in two ways:

- in rivers, the total shear-stress τ_0 is often prescribed as a function of the skin friction τ_{0s}. This approach, known as the 'alluvial friction' method, relies on an assumption that the bedforms are in equilibrium with the flow, and that their frictional characteristics are uniquely determined by the bulk flow.
- a two-stage approach, in which the height and wavelength of the sandwaves are determined first, and these are then used to determine the friction.

The second method is more appropriate in the sea, because the tidal currents and the surface waves vary rapidly with a time-scale of hours, whereas the sandwaves respond rather slowly with a time-scale of days, so that the flow is generally not in equilibrium with the bedforms. If measured values of the height and wavelength of the sandwaves are available, then these can be used directly to give a more accurate calculation of the friction.

An early but simple alluvial friction method is that given by Engelund (1966):

$$\theta = 2{\cdot}5(\theta_s - 0{\cdot}06)^{1/2} \qquad\qquad\qquad \text{SC (94)}$$

where $\theta = \tau_0/g\rho(s-1)d$
$\qquad \theta_s = \tau_{0s}/g\rho(s-1)d$
$\qquad \tau_0 = $ total shear-stress
$\qquad \tau_{0s} = $ skin-friction shear-stress
$\qquad g = $ acceleration due to gravity
$\qquad \rho = $ water density
$\qquad s = $ relative density of sediment
$\qquad d = $ grain diameter

Another well-established alluvial friction method is that of White *et al.* (1980). The method is complicated, and the reader is referred either to the original paper or to Fisher (1993), where a step-by-step procedure and a worked example are given.

An example of the two-stage method is that of Van Rijn (1984). The height Δ_s and wavelength λ_s of dunes are predicted by using Equations (83a–d). These are then used to obtain k_s from the following equation:

$$k_s = 1{\cdot}1\Delta_s[1 - \exp(-25\Delta_s/\lambda_s)] + 3d_{90} \qquad\qquad \text{SC (95)}$$

This value of k_s is then used in the Chézy formula (similar to Equation (37)) to obtain the total shear-stress τ_0.

Further details and examples of the above methods are given by Fisher (1993).

Procedure

1. *Example 7.4. Total shear-stress for current*

To calculate the total shear stress τ_0 produced by a tidal current over a rippled sea bed with $d_{50} = 200\ \mu\text{m}$ in sea water at 10 °C and 35 ppt:

Estimate wavelength of ripples (Equation (81a))	λ_r	0·20 m
Estimate height of ripples (Equation (81b))	Δ_r	0·0286 m
Calculate from Equation (90), $z_0 = 1 \times (0·0286)^2/0·20$		0·00408 m
Obtain water depth	h	10 m
Use Equation (37) to obtain		
$C_D =$		
$\{0·40/[1 + \ln(0·00408/10)]\}^2 =$		0·00346
Obtain depth-averaged velocity	$\bar{U}$	0·50 m s^{-1}
Use Equation (30) to calculate total shear-stress		
$\tau_0 = 1027 \times 0·00346 \times 0·50^2 =$		0·887 N m^{-2}

2. When applied to the numerical values used in Example 7.4, with a quartz sand bed having $d_{35} = 175$ μm, $d_{50} = 200$ μm, $d_{65} = 230$ μm, $d_{90} = 313$ μm, the river-based methods give the following predictions for the total shear-stress:

Engelund $\tau_0 = 0·965$ N m^{-2}
White *et al.* $\tau_0 = 0·384$ N m^{-2}
Van Rijn $\tau_0 = 0·600$ N m^{-2}

These methods predict dunes rather than ripples as the main roughness element.

3. *Example 7.5. Total shear-stress for waves*

To calculate the total shear-stress amplitude τ_w produced by waves over a rippled sea bed, using the same inputs as for Example 7.3:

Calculate ripple height and wavelength as in Example 7.3. Nielsen's method gave:		
ripple height	Δ_r	0·0452 m
ripple wavelength	λ_r	0·265 m
Nielsen's method for calculating z_{0f} uses Equation (90) with $a_r = 0·267$:	z_{0f}	$2·06 \times 10^{-3}$ m
Calculate the sediment-transport component of z_0 from Equation (92)	z_{0t}	$4·14 \times 10^{-4}$ m
Calculate the skin-friction component of z_0 from Equation (25)	z_{0s}	$1·67 \times 10^{-5}$ m
Calculate total z_0 using Equation (43)	z_0	$2·49 \times 10^{-3}$ m
Calculate the wave friction factor by Swart's method, Equation (60)	f_{wr}	0·139
Calculate the amplitude of the total bed shear-stress using Equation (57)	τ_w	$6·84$ N m^{-2}

For comparison, the method of Raudkivi gives $\tau_w = 11·0$ N m^{-2}, and the method of Grant and Madsen gives $\tau_w = 10·8$ N m^{-2}.

4. Details of the methods available in SandCalc for total wave bed shear-stress are given in Section 4.6.

Suspended sediment

8

8. *Suspended sediment*

8.1. GENERAL

For current speeds or wave conditions significantly above the threshold of motion, sand is entrained off the bed and into suspension, where it is carried at the same speed as the current. When this happens, the proportion of sediment carried in suspension is generally much larger than that being carried simultaneously as bedload, and hence the *suspended load* is an important contribution to the total sediment transport rate.

An important factor in the design of cooling water intakes for power stations is prevention of ingress of suspended sediment, for which calculations of the concentrations and grain sizes at the height of the intake are required.

8.2. SUSPENSION CRITERIA AND GRAIN SIZE

Knowledge

For grains to remain in suspension, their settling velocity must be smaller than the upward turbulent component of velocity, which is related to u_*. This leads to a criterion for the threshold of suspension of sediment given approximately by the relationship

$$u_{*s} = w_s \tag{96}$$

where

u_{*s} = skin-friction velocity
w_s = settling velocity of grains (see Section 8.3)

For *mixed sediment*, Equation (96) can be applied to each grain-size fraction. If the bed material is widely graded, only the finer fractions are suspended, with the coarser ones moving as bedload. The best procedure in this case is to divide the sediment into a number of grain-size classes, each comprising a narrow band of grain diameters, and treat each class separately. A simpler, but less rigorous, approach is to select a single grain size which is representative of the whole sample.

Ackers and White (1973) found that d_{35} of the bed sediment gave the best predictions of total sediment transport rate in rivers.

Van Rijn (1984) related the median suspended grain diameter $d_{50,s}$ to the median bed grain diameter $d_{50,b}$ through a sorting parameter $\sigma_s = 0 \cdot 5(d_{84}/d_{50} + d_{50}/d_{16})$ and the transport parameter $T_s = (\tau_{0s} - \tau_{cr})/\tau_{cr}$ via the relation:

$$d_{50,s}/d_{50,b} = 1 + 0 \cdot 011(\sigma_s - 1)(T_s - 25)$$

$$\text{for } 0 < T_s < 25 \quad \text{SC (97a)}$$

$$= 1 \qquad \qquad \text{for } T_s \geq 25 \qquad \text{SC (97b)}$$

Equation (97) is only valid if $(\sigma_s - 1) < [0 \cdot 011(25 - T_s)]^{-1}$; otherwise it predicts $d_{50,s} < 0$.

Fredsøe and Deigaard (1992) exclude all grains for which $w_s > 0 \cdot 8 u_{*s}$ from the suspension process, and take the median diameter of the remainder as the representative grain size in suspension.

Whitehouse (1995) found from field measurements that the median grain diameter in suspension $d_{50,s}$ corresponded to values of between d_2 and d_{15} of the bed sediment (coarser for stronger flows), with d_{10} being a typical value.

8.3. SETTLING VELOCITY

Knowledge

The *settling velocity* (or fall velocity, or terminal velocity) of sand grains in water is determined by their diameter and density, and the viscosity of the water. At the finest end of the sand range of diameters ($d = 62 \, \mu\text{m}$), grains settle according to Stokes' law of viscous drag; at the coarsest end ($d = 2 \, \text{mm}$) they obey a quadratic bluff-body drag law; and intermediate sizes experience a mixture of viscous and bluff-body drag. The drag on natural irregularly-shaped sand grains is simpler than that for spheres, because the angular surfaces and variations in shape between grains tend to cause a more gradual flow separation process. Consequently, it is better not to treat sand grains as spheres for such calculations.

There are a number of formulae for calculating the settling velocity w_s of isolated sand grains in still water. Some of these require the dimensionless grain size D_* to be calculated

$$D_* = \left[\frac{g(s-1)}{\nu^2}\right]^{1/3} d \qquad (98)$$

where g = acceleration due to gravity = $9\cdot81$ m s^{-2}
ν = kinematic viscosity of water
d = median sieve diameter of grains
s = ratio of densities of grain and water

The formula of Gibbs *et al.* (1971) for spheres,

$$w_s = \frac{[9\nu^2 + gd^2(s-1)(3\cdot869\times10^{-3}+0\cdot024801d)]^{1/2}-3\nu}{0\cdot011607+0\cdot074405d} \qquad (99)$$

in which all units must be in cgs, has in the past been widely (but inappropriately) used for sands.

The formula of Hallermeier (1981) for natural sand is

$$w_s = \frac{\nu D_*^3}{18d} \qquad \text{for } D_*^3 \leq 39 \qquad \text{SC}(100a)$$

$$w_s = \nu \frac{D_*^{2\cdot1}}{6d} \qquad \text{for } 39 < D_*^3 < 10^4 \qquad \text{SC}(100b)$$

$$w_s = \frac{1\cdot05\nu D_*^{1\cdot5}}{d} \qquad \text{for } 10^4 \leq D_*^3 < 3\times10^6 \qquad \text{SC}(100c)$$

The formula of Van Rijn (1984) for natural sand is

$$w_s = \frac{\nu D_*^3}{18d} \qquad \text{for } D_*^3 \leq 16\cdot187 \qquad \text{SC}(101a)$$

$$w_s = \frac{10\nu}{d}[(1+0\cdot01D_*^3)^{1/2}-1]$$

$$\text{for } 16\cdot187 < D_*^3 \leq 16187 \quad \text{SC}(101b)$$

$$w_s = \frac{1\cdot1\nu D_*^{1\cdot5}}{d} \qquad \text{for } D_*^3 > 16187 \qquad \text{SC}(101c)$$

Zanke (1977) also gave Equation (101b).

Soulsby derived the following formula for natural sands, based on optimising two coefficients in a combined viscous plus bluff-body drag law against data for irregular grains:

$$w_s = \frac{\nu}{d}[(10\cdot36^2 + 1\cdot049D_*^3)^{1/2} - 10\cdot36] \text{ for all } D_* \quad \text{SC (102)}$$

A comparison of the predictions of settling velocity as given by these four formulae (Equations (99)–(102)) has been carried out against a large data set comprising 115 measurements of settling velocities of natural sands and irregular shaped lightweight grains. The data were collated and tabulated by Hallermeier (1981).

Table 11 shows the percentage of predictions lying within 10% or 20% of the observations.

Equation (102) gave the best results, and is also the simplest. This good agreement may partly be because the coefficients were optimised for the test data set. The Hallermeier and Van Rijn formulae were almost as good, but are more complicated, so Equation (102) is recommended. The poor performance of the Gibbs formula is because it was intended for spheres, not natural grains. Figure 26 shows a non-dimensional plot of $w_s/[g(s-1)\nu]^{1/3}$, versus D_* for this data set, and the curve corresponding to Equation (102). Figure 27 shows curves of w_s versus d for the case $g = 9\cdot81 \text{ m s}^{-2}$, $\rho_s = 2650 \text{ kg m}^{-3}$, temperature $= 10\,^{\circ}\text{C}$, salinity $= 35$ ppt, typical of sand in seawater, and for the case of sand in fresh water at $20\,^{\circ}\text{C}$.

At high concentrations the flows around adjacent settling grains interact to give a larger drag than for the same grain in isolation (hindered settling). This causes the hindered settling velocity w_{sC} at high concentration to be smaller than that at low concentration, w_s. Applying similar reasoning to that which led to Equation (102), but now including a factor $(1-C)^{-4\cdot7}$ in the

Table 11. *Comparison of predictions of settling velocity*

Formula	Equation	10%	20%
Gibbs *et al.*	99	35	50
Hallermeier	100	60	89
Van Rijn	101	59	90
Soulsby	102	66	90

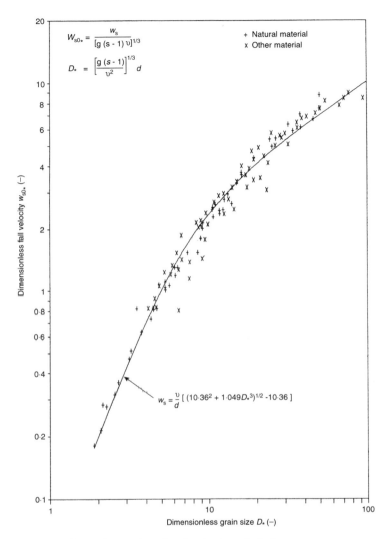

Figure 26. *Settling velocity of sand grains (universal plot)*

grain drag formula (obtained empirically by Wen and Yu, 1966), yields the following formula for the settling velocity w_{sC} of grains in a dense suspension having concentration C:

$$w_{sC} = \frac{\nu}{d}\{[10{\cdot}36^2 + 1{\cdot}049(1 - C)^{4{\cdot}7}D_*^3]^{1/2} - 10{\cdot}36\}$$

$$\text{for all } D_* \text{ and } C. \quad (103)$$

Equation (103) reduces to Equation (102) for $C \rightarrow 0$, and is compatible with Equation (18) for fluidisation.

For small values of D_*, Equation (103) shows that the ratio $w_{sC}/w_s = (1 - C)^{4.7}$, whereas for large values of D_*, Equation (103) gives $w_{sC}/w_s = (1 - C)^{2.35}$. In practice, the effect of hindered settling need only be taken into account for concentrations larger than 0·05, which usually occur only within a few millimetres of the bed, since the difference between w_s and w_{sC} is less than 20% for smaller concentrations.

Procedure

1. To calculate the settling velocity of grains at low concentration, obtain the median grain size d_{50} of a sand sample (preferably taken from suspension) by sieve analysis (Section 2.2).

    If the temperature and salinity are close to 10 °C and 35 ppt, respectively, read w_s for the given $d = d_{50}$ from Figure 27 or, for greater accuracy:

    Example 8.1. Settling velocity

Obtain temperature in degrees C		10
Obtain salinity in ppt		35
Calculate kinematic viscosity (Example 2.1)	ν	1.36×10^{-6} m^2 s^{-1}
Obtain grain density	ρ_s	2650 kg m^{-3}
Calculate water density (Example 2.1)	ρ	1027 kg m^{-3}
Calculate $s = \rho_s/\rho$		2·58
Obtain grain diameter	d	0·2 mm
Calculate D_* from Equation (98)	D_*	4·06
Calculate settling velocity from Equation (102)	w_s	0·0202 m s^{-1}

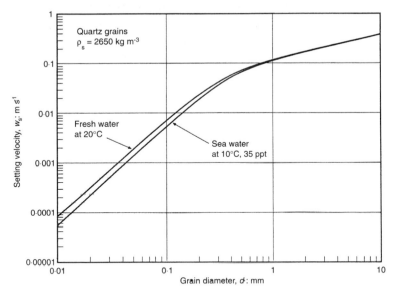

Figure 27. Settling velocity of quartz grains of sieve diameter d at low concentration in still water

For comparison, the value of w_s predicted by Van Rijn's method (Equation (101)) is $0.0199 \, \mathrm{m \, s^{-1}}$ and by Hallermeier's method (Equation (100)) is $0.0216 \, \mathrm{m \, s^{-1}}$.

2. If volume concentration is greater than 0.05, instead use Equation (103).

8.4. CONCENTRATION UNDER CURRENTS

Knowledge

In a sand suspension the settling of the grains towards the bed is counterbalanced by diffusion of sand upwards due to the turbulent water motions (including vertical components of velocity) near the bed. The equation governing this balance is

$$w_s C = -K_s \frac{dC}{dz} \tag{104}$$

where w_s = settling velocity of sediment grains
C = volume concentration of sediment at height z
K_s = eddy diffusivity of sediment

The eddy diffusivity depends on the turbulence in the flow and on the height above the bed. Equation (104) can be solved to give the vertical distribution of the concentration of the suspended sediment, subject to certain assumptions. Different assumptions about the eddy diffusivity of the sediment lead to different expressions for the concentration profile. The shape of the profile depends on the ratio

$$b = \frac{w_s}{\kappa u_*} \tag{105}$$

where b = Rouse number, or suspension parameter
κ = von Karman's constant $= 0.40$
u_* = total friction velocity

Note that the *entrainment* of sediment from the bed is governed by the skin-friction $\tau_{0s}(= \rho u_{*s}^2)$, whereas the *diffusion* of the sediment higher into the water column is governed by the total shear-stress $\tau_0(= \rho u_*^2)$. This is because the form drag of ripples does not act directly on grains lying on the surface of the bed, but it generates turbulence which governs the diffusion process. The distinction disappears for sheet-flow conditions (see Section 7.2), where $u_* = u_{*s}$.

Sediment suspension by a *current* in the sea is similar to that found in rivers.

If the eddy diffusivity is assumed to increase linearly with height above the bed ($K_s = \kappa u_* z$), the corresponding concentration profile is the *power-law* profile:

$$C(z) = C_a \left(\frac{z}{z_a}\right)^{-b} \qquad \text{SC (106)}$$

If the eddy diffusivity is assumed to vary parabolically with height ($K_s = \kappa u_* z[1 - (z/h)]$), the *Rouse* profile is obtained:

$$C(z) = C_a \left[\frac{z}{z_a} \cdot \frac{h - z_a}{h - z}\right]^{-b} \qquad \text{SC (107)}$$

If the eddy diffusivity is assumed to vary parabolically with height in the lower half, and be constant with height in the upper half, of the water column, the Van Rijn (1984) profile is obtained.

Note that Van Rijn also took account of the difference between diffusion of sediment and fluid momentum, and the density stratification by the sediment, by introducing a modified form of the exponent b:

$$C(z) = C_a \left[\frac{z}{z_a} \cdot \frac{(h - z_a)}{(h - z)} \right]^{-b'} \quad \text{for } z_a < z < \frac{h}{2} \qquad \text{SC (108a)}$$

$$C(z) = C_a \left(\frac{z_a}{h - z_a} \right)^{b'} \exp\left[- 4b' \left(\frac{z}{h} - \frac{1}{2} \right) \right] \text{ for } \frac{h}{2} < z < h$$

$$\text{SC (108b)}$$

with

$$b' = \frac{b}{B_1} + B_2 \qquad \text{SC (108c)}$$

$$B_1 = 1 + 2 \left(\frac{w_s}{u_*} \right)^2 \quad \text{for } 0{\cdot}1 < w_s/u_* < 1 \qquad \text{SC (108d)}$$

$$= 2 \qquad \qquad \text{for } w_s/u_* \geq 1$$

$$B_2 = 2{\cdot}5 \left(\frac{w_s}{u_*} \right)^{0{\cdot}8} \left(\frac{C_a}{0{\cdot}65} \right)^{0{\cdot}4} \quad \text{for } 0{\cdot}01 \leq \frac{w_s}{u_*} \leq 1$$

$$\text{SC (108e)}$$

$$= 0 \quad \text{for } w_s > u_* \quad \text{or} \quad z_a > 0{\cdot}1h$$

In Equations (106) to (108):

$z =$ height above seabed

$z_a =$ a reference height near the seabed

$C(z) =$ sediment concentration at height z

$C_a =$ sediment reference concentration at height z_a

$h =$ water depth

$b =$ Rouse number (Equation (105)).

The concentration can be expressed as volume/volume or mass/volume, provided that $C(z)$ and C_a have the same units (see Section 2.3).

A comparison of the shapes of the three profiles for the case $b = 1$, with $z_a/h = 0.01$, is shown in Figure 28a. The Rouse profile is the most widely used, especially in rivers. Examples of the Rouse profile for various values of b are shown in Figure 28b. For fine grains and strong currents (small b) the sediment is well mixed throughout the water column, whereas for coarse grains and weak currents (large b) the sediment is concentrated mainly near the bed.

The Rouse profile is less well-suited for use in the sea because the parabolic eddy diffusivity reduces to zero at the surface, resulting in a prediction of zero sediment concentration at the water surface. This is contrary to observations, especially if waves are present. The power-law and Van Rijn profiles both have non-zero diffusivity at the surface, and are therefore more suitable.

The power-law has the attraction of simplicity for mathematical manipulation, especially since its greatest accuracy occurs in the lower part of the water column, where concentrations are largest.

The Van Rijn profile probably corresponds best to data, and is recommended for use with currents alone in the sea.

The *reference concentration* C_a, and reference height z_a, must be specified in order for Equations (106)–(108) to give usable predictions of concentration. Several expressions for these exist. Garcia and Parker (1991) tested seven of these against a large data set, and concluded that the best two were:

Smith and McLean (1977)

$$C_a = \frac{0.00156 T_s}{1 + 0.0024 T_s}$$

at height $z_a = \dfrac{26.3 \tau_{cr} T_s}{\rho g(s-1)} + \dfrac{d_{50}}{12}$ \hfill SC (109)

Van Rijn (1984)

$$C_a = \frac{0.015 d T_s^{3/2}}{z_a D_*^{0.3}}$$ \hfill SC (110)

at height $z_a = \Delta_s/2$, with Δ_s given by Equation (83) and minimum value of $z_a = 0.01h$.

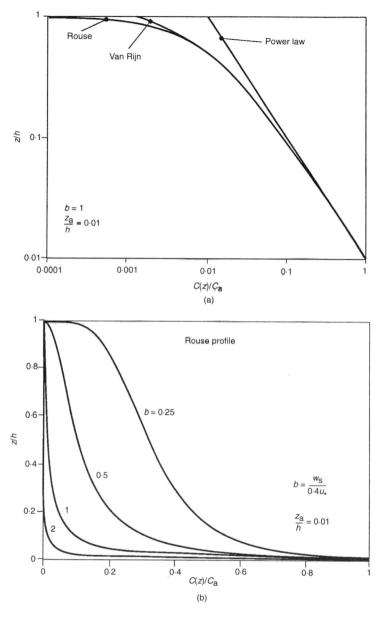

Figure 28. Suspended sediment concentration profiles: (a) comparison of three formulae (logarithmic axes); (b) variation with grain size and current speed (linear axes)

A more recent expression which also seems to give good results is that of Zyserman and Fredsøe (1994)

$$C_a = \frac{0\cdot331(\theta_s - 0\cdot045)^{1\cdot75}}{1 + 0\cdot720(\theta_s - 0\cdot045)^{1\cdot75}} \qquad \text{SC (111)}$$

at height $z_a = 2d_{50}$.

In Equations (109)–(111):

C_a = concentration (volume/volume) at height z_a

z_a = reference height

τ_{0s} = skin-friction bed shear-stress

τ_{cr} = threshold bed shear-stress for motion of sediment

$T_s = (\tau_{0s} - \tau_{cr})/\tau_{cr}$

d_{50} = median grain diameter

h = water depth

g = acceleration due to gravity

ρ = density of water

ρ_s = density of sediment material

$s = \rho_s/\rho$

ν = kinematic viscosity of water

$$D_* = \left[\frac{g(s-1)}{\nu^2}\right]^{1/3} d_{50}$$

$$\theta_s = \frac{\tau_{0s}}{g\rho(s-1)d_{50}} = \text{skin-friction Shields parameter}$$

Δ_s = height of sandwaves

Procedure

1. To calculate the suspended sediment concentration at 1 m above the bed, using the power-law concentration profile with Smith and McLean's reference concentration for a steady current (tidal and wind-driven currents can be regarded as steady), for quartz grains in sea-water at 10 °C, 35 ppt, obtain values of:

Example 8.2. Concentration under currents

depth-averaged current speed	$\bar{U}$	$1 \cdot 0$ m s^{-1}
water depth	h	10 m
grain-size distribution of bed	d_{10}	$0 \cdot 17$ mm
	d_{50}	$0 \cdot 20$ mm
	d_{90}	$0 \cdot 23$ mm

In this example, the sediment is very well-sorted, and d_{50} can be taken as the representative grain size, d.

Calculate settling velocity of representative grains (see Example 8.1). For $d = 0 \cdot 20$ mm:	w_s	$0 \cdot 0202$ m s^{-1}
Calculate skin-friction shear-stress (see Example 3.3), e.g. use Soulsby formula	τ_{0s}	$0 \cdot 952$ m s^{-1}
Check if bed is rippled by calculating Shields' parameter (see Equation (2a))	θ_s	$0 \cdot 299$
If $\theta_s < 0 \cdot 8$, bed will be rippled. (Also check if there are sandwaves—in example, assume none)		Rippled
Calculate total drag coefficient C_D above rippled bed, using Equation (37) with $z_0 = 0 \cdot 006$ m (Table 7 for rippled sand)	C_D	$0 \cdot 00388$
Calculate total skin friction $u_* = C_D^{\frac{1}{2}} \bar{U}$	u_*	$0 \cdot 0623$ m s^{-1}
Calculate threshold shear-stress (see Example 6.3)	τ_{cr}	$0 \cdot 176$ m s^{-1}
Calculate $T_s = \dfrac{\tau_{0s} - \tau_{cr}}{\tau_{cr}}$	T_s	$4 \cdot 41$

Calculate reference height and reference concentration from Equation (109)

$$\frac{26\cdot3 \times 0\cdot176 \times 4\cdot41}{1027 \times 9\cdot81(2\cdot58 - 1)}$$

$$+\frac{0\cdot2 \times 10^{-3}}{12} = \qquad z_a \qquad 1\cdot30 \times 10^{-3}\,\text{m}$$

$$\frac{0\cdot00156 \times 4\cdot41}{1 + 0\cdot0024 \times 4\cdot41} = \qquad C_a \qquad 6\cdot81 \times 10^{-3}$$

Calculate Rouse number

$$\frac{w_s}{\kappa u_*} = \frac{0\cdot0202}{0\cdot40 \times 0\cdot0623} = \qquad b \qquad 0\cdot811$$

Select height at which concentration is required $\qquad z \qquad 1\cdot0\,\text{m}$

Calculate concentration at height z by using Equation (106)

$$6\cdot81 \times 10^{-3}$$

$$\times \left(\frac{1\cdot0}{1\cdot30 \times 10^{-3}}\right)^{-0\cdot811} = \qquad C(z) \qquad 3\cdot1 \times 10^{-5}$$

If preferred convert to $\text{kg}\,\text{m}^{-3}$ by multiplying by $2650\,\text{kg}\,\text{m}^3$ (see Table 4) $\qquad C_M(z) \qquad 0\cdot082\,\text{kg}\,\text{m}^{-3}$

For comparison:

using Van Rijn profile and Van Rijn C_a, z_a $\qquad C_M(z) \qquad 0\cdot0616\,\text{kg}\,\text{m}^{-3}$

using Rouse profile and Zyserman and Fredsøe $\qquad C_a, z_a\; C_M(z) \qquad 0\cdot121\,\text{kg}\,\text{m}^{-3}$

The different methods give answers for C_M at $z = 1\,\text{m}$ varying by a factor of 2. The median suspended grain diameter could be calculated using the Van Rijn method (Equation (97)) to give $d_{50,s} = 0\cdot194\,\text{mm}$, for which the concentration at $z = 1\,\text{m}$ in Example 8.2 for the power-law profile method becomes $C_M(z) = 0\cdot104\,\text{kg}\,\text{m}^{-3}$.

8.5. CONCENTRATION UNDER WAVES

Knowledge

Under *waves*, the suspended sediment is confined to the relatively thin (a few mm or cm) wave boundary layer (see Section 4.4). For a rippled bed, the eddy diffusivity (see Section 8.4) is constant with height, and the concentration profile is given by

$$C(z) = C_0 e^{-z/l} \qquad\qquad \text{SC (112)}$$

where z = height above sea bed
$\quad C(z)$ = sediment concentration (volume/volume)
$\qquad\qquad$ at height z
$\quad C_0$ = reference concentration (volume/volume)
$\qquad\qquad$ at the sea bed $(z = 0)$
$\quad l$ = decay length scale

Various expressions have been given for l and C_0, of which one of the most widely used is that of Nielsen (1992) for rippled beds:

$$l = 0.075 \frac{U_w}{w_s} \Delta_r \quad \text{for} \quad \frac{U_w}{w_s} < 18 \qquad\qquad \text{SC (113}a\text{)}$$

$$l = 1.4\Delta_r \qquad\qquad \text{for} \quad \frac{U_w}{w_s} \geq 18 \qquad\qquad \text{SC (113}b\text{)}$$

$$C_0 = 0.0050\theta_r^3 \qquad\qquad \text{SC (113}c\text{)}$$

$$\theta_r = \frac{f_{wr} U_w^2}{2(s-1)gd(1 - \pi\Delta_r/\lambda_r)^2} \qquad\qquad \text{SC (113}d\text{)}$$

where U_w = wave orbital velocity amplitude (see Section 4.4)
$\quad w_s$ = grain settling velocity (see Section 8.3)
$\quad \Delta_r$ = ripple height
$\quad \lambda_r$ = ripple wavelength
$\quad f_{wr}$ = rough-bed wave friction factor (Equation (60b))
$\qquad\quad = 0.00251 \exp(5.21 r^{-0.19})$

$$r = \frac{U_w T}{5\pi d_{50}}$$

$\quad d_{50}$ = median grain diameter
$\quad T$ = wave period
$\quad s$ = relative density of sediment
$\quad g$ = acceleration due to gravity

Equation (112) is also sometimes used for sheet-flow (flat-bed) conditions, although recent evidence indicates that a better expression in this case is analogous to Equation (106) for currents:

$$C(z) = C_a \left(\frac{z}{z_a}\right)^{-b} \qquad\qquad\text{SC (114)}$$

Experimental values of b do not necessarily follow theoretical expectations, and are typically $b = 1{\cdot}7$ for $d_{50} = 0{\cdot}13$ mm and $b = 2{\cdot}1$ for $d_{50} = 0{\cdot}21$ mm, for all wave conditions within the ranges $0{\cdot}4 < U_w < 1{\cdot}3$ m s^{-1} and $5 < T < 12s$. If $z_a = 1$cm, values of $C_a \simeq 4 \times 10^{-4}$ for $U_w = 0{\cdot}5$ m s^{-1} and $C_a \simeq 8 \times 10^{-3}$ for $U_w = 1{\cdot}3$ m s^{-1} were found experimentally for $d_{50} = 0{\cdot}21$ mm (Ribberink and Al-Salem, 1991).

With the above caveat, Equation (114) can be used with the theoretical value $b = w_s/(\kappa u_{*w})$, and with C_a and z_a obtained from the method of Zyserman and Fredsøe (Equation (111)), with θ_{ws} replacing θ_s, for example.

Procedure

1. *Example 8.3. Concentration under waves*

   To calculate the concentration of sediment suspended by waves, obtain values of:

height at which concentration is required	z	0·10 m
wave orbital velocity amplitude	U_w	0·310 m s^{-1}
wave period	T	6 s
median grain diameter	d_{50}	0·2 mm
relative density of sediment	s	2·58

   (The example values are the same as in Example 7.3.)

Calculate value of $U_w T/(5\pi d_{50})$	r	592

Calculate wave friction factor
(Equation (60b))　　　　　　　　f_{wr}　　　　0·0118

Calculate amplitude of
grain-related Shields parameter　θ_{ws}　　　0·183

If $\theta_{ws} < 0·8$, then the bed is rippled
and Equations (112) and (113)
should be used.

If $\theta_{ws} > 0·8$, then the bed is flat
with sheet-flow and Equation (114)
should be used.

In the example, $\theta_{ws} < 0·8$, so the bed
is rippled, and Example 7.3 gives:　Δ_r　　0·0452 m

　　　　　　　　　　　　　　　　λ_r　　0·265 m

Calculate from Equation (113d)　　θ_r　　0·849

Calculate w_s (see Example 8.1)　　w_s　　0·0202 m s^{-1}

Calculate U_w/w_s　　　　　　　　　　　　49·5

Calculate l from Equation (113b)　l　　0·0633 m

Calculate concentration at z from
Equation (112)　　　　　　　　　$C(z)$　　$4·48 \times 10^{-4}$

or as mass per volume $= \rho_s C(z)$　$C_M(z)$　1·19 kg m^{-3}

2. For comparison, if a sheet flow over the ripples is assumed, then Equation (114), with b calculated using the *total* u_* (including form drag of the ripples), and using the Zyserman and Fredsøe reference concentration, yields $C_M(z) = 0·836\,\mathrm{kg\,m^{-3}}$ at $z = 0·10\,\mathrm{m}$.

3. Calculations of concentration profiles are very sensitive to the settling velocity, and hence to the suspended grain size. In Example 8.3, a uniform bed material was assumed, and hence $d_{50,s} = d_{50,b}$. In reality, $d_{50,s}$ will be appreciably smaller than $d_{50,b}$, but reliable methods of predicting this under waves have not yet been established.

8.6. CONCENTRATION UNDER WAVES PLUS CURRENTS

Knowledge

Under *combined waves and currents*, sediment is suspended within the wave boundary layer, and diffused further up into the flow by the turbulence associated with the current. Both these processes are affected by the interaction between wave and current boundary layers discussed in Section 5.3. The form of the concentration profile in these conditions is not well established, but is typically of the following form:

$$C(z) = C_a \left(\frac{z}{z_a} \right)^{-b_{max}} \qquad \text{for } z_a \leq z \leq z_w \qquad \text{SC (115a)}$$

$$C(z) = C(z_w) \left(\frac{z}{z_w} \right)^{-b_m} \qquad \text{for } z_w < z \leq h \qquad \text{SC (115b)}$$

with

$$b_{max} = \frac{w_s}{\kappa u_{*max}} \qquad \text{SC (115c)}$$

$$b_m = \frac{w_s}{\kappa u_{*m}} \qquad \text{SC (115d)}$$

$$z_w = \frac{u_{*max} T}{2\pi} = \text{wave boundary thickness} \qquad \text{SC (115e)}$$

and

$$z = \text{height above sea bed}$$

$$C(z) = \text{sediment concentration at height } z$$

$$C_a = \text{reference concentration at height } z_a$$

$$z_a = \text{reference height near sea bed}$$

$$C(z_w) = \text{concentration at } z = z_w \text{ calculated from}$$
$$\text{Equation (115a)}$$

$$w_s = \text{sediment settling velocity}$$

$$\kappa = \text{von Karman's constant} = 0{\cdot}40$$

$$u_{*max} = (\tau_{max}/\rho)^{1/2}$$

$$u_{*m} = (\tau_m/\rho)^{1/2}$$

τ_{max} = maximum bed shear-stress in wave cycle

τ_m = mean bed shear-stress in wave cycle

T = wave period.

The reference concentration C_a can be calculated from one of the expressions given in Section 8.4 (Equations (109)–(111)). For example, the Zyserman and Fredsøe (1994) expression can be used, with $\theta_{max,s}$ instead of θ_s.

The above method is designed for flat-bed (sheet-flow) conditions, so that the appropriate roughness length to use when calculating τ_m and τ_{max} is $z_0 = d_{50}/12$. These conditions apply provided that $\tau_{max,s} > 0{\cdot}8g(\rho_s - \rho)d_{50}$ (i.e. $\theta_{max,s} > 0{\cdot}8$).

If $\theta_{max,s} < 0{\cdot}8$, then the bed is likely to be rippled (but note that this criterion is not firmly established for wave-plus-current flows). For a rippled bed, skin-friction values of τ_m and τ_{max} should be used in the calculation of the reference concentration, C_a, but total-stress values should be used to calculate b_{max} and b_m.

Van Rijn (1993) extended his current-only expression for the concentration profile (Equation (108)) to include the effects of waves. An eddy diffusivity profile due to the waves was proposed and added in a sum-of-squares fashion to the diffusivity due to a current. The Van Rijn expression for the reference concentration (Equation (110)) was adapted to include wave-induced bed shear-stresses, and the effect of ripples was also adapted to account for waves. The method applies to both rippled-bed and sheet-flow conditions. The resulting computations for the concentration profile (and sediment transport rate) are performed by Van Rijn's program TRANSPOR, which is included with his book (Van Rijn, 1993), although it should be noted that the user must exercise some judgement to select the suspended grain size, and current-related and wave-related bed roughnesses, used as input values.

The most advanced method of calculating suspended sediment concentration profiles (and transport rates) is by means of a numerical model of the wave-plus-current boundary-layer, utilising some form of turbulent energy closure. Equations for momentum, turbulence properties and sediment concentration are solved at a grid of points in the vertical, at each time step in a wave cycle. Once the model has converged to a periodic solution, the profiles of the instantaneous and time-mean horizontal velocity and sediment concentration through the wave cycle are obtained. Davies *et al.* (1997) describe four models of this type, developed by A. G. Davies, J. S. Ribberink, A. Temperville and the Danish 'STP' model, each with a different form of turbulence closure and different assumptions about the sediment behaviour (form of reference concentration, turbulence damping, graded sediments). The models were compared with measurements of suspended sediment concentration up to a height of 0·10 m made in an oscillating water tunnel (OWT) with steady-plus-oscillatory velocities over a bed of sand with $d_{50} = 0·21$ mm. The Davies, Temperville and STP models gave agreement with the measured time-mean concentration profile generally to within a factor 3. The Davies, Ribberink and Temperville models are restricted to sheet-flow, flat-bed conditions (as occurred in the OWT measurements), while the STP model can also deal with a rippled bed. Future improvements in prediction capability are likely to lie in further development of this type of model.

Many aspects of sediment response and suspension under combined waves and currents are still poorly understood, so results should be treated with caution. The results are strongly dependent on the methods used, the assumptions made, and on input parameters such as the suspended grain size and the water temperature. A sensitivity analysis involving all of these is recommended when making predictions for practical applications.

Procedure

1. The procedure to calculate the concentration of sediment suspended by combined waves and currents is illustrated by combining the steady current input values used in Example 8.2 collinearly ($\phi = 0°$) with the wave input values used in Example 8.3. See these examples for some of the steps:

Example 8.4. Concentration under waves and currents

Obtain values of:

water depth	h	10 m
water density (10 °C, 35 ppt)	ρ	1027 kg m^{-3}
median grain diameter	d_{50}	0·2 mm

density of sediment grains	ρ_s	2650 kg m^{-3}
depth-averaged current speed	$\bar{U}$	1·0 m s^{-1}
wave orbital velocity amplitude	U_w	0·310 m s^{-1}
wave period	T	6 s
angle between directions of current and wave	ϕ	0°
Calculate settling velocity	w_s	0·0202 m s^{-1}
Calculate skin-friction bed shear-stresses, using the DATA13 method (see Example 5.1) with grain-related roughness	$z_{0s} = d_{50}/12$	0·0167 mm
mean skin-friction bed shear-stress	τ_{ms}	2·53 N m^{-2}
maximum skin-friction bed shear-stress	$\tau_{max,s}$	2·95 N m^{-2}
mean skin-friction velocity $= (\tau_{ms}/\rho)^{1/2}$	u_{*ms}	0·0496 m s^{-1}
maximum skin-friction velocity $= (\tau_{max,s}/\rho)^{1/2}$	$u_{*max,s}$	0·0536 m s^{-1}
mean skin-friction Shields parameter (Equation (2a))	θ_{ms}	0·794
maximum skin-friction Shields parameter (Equation (2a))	$\theta_{max,s}$	0·927

Since $\theta_{max,s} > 0·8$ (and even $\theta_{ms} \simeq 0·8$), the bed is expected to be flat with sheet flow. The total bed shear-stresses, etc. are therefore equal to the skin-friction values.

Calculate reference concentration using method of Zyserman and Fredsøe (Equation (111)), with θ_s replaced by $\theta_{max,s}$:	C_a	0·168

Calculate reference height $= 2d_{50}$	z_a	0.4×10^{-3} m
Calculate b_{max} (Equation (115c))	b_{max}	0.942
Calculate b_m (Equation (115d))	b_m	1.02
Calculate wave boundary layer thickness (Equation (115e))	z_w	0.0512 m
Calculate $C(z_w)$ (Equation (115a) with $z = z_w$)	$C(z_w)$	1.74×10^{-3}
Select height at which concentration is required z		0.10 m
Calculate concentration at height z (Equation (115b)) $C(z)$		8.82×10^{-4}
Convert to mass concentration $= \rho_s C(z)$	$C_M(z)$	2.34 kg m^{-3}

The value $C_M(z) = 2.34 \, \text{kg m}^{-3}$ at $z = 0.10$ m compares with the value $1.19 \, \text{kg m}^{-3}$ found at the same height in Example 8.3 for the same waves without a current. A similar calculation using $z = 1.0$ m in Example 8.4 gives $C_M(z) = 0.225 \, \text{kg m}^{-3}$, which compares with the value $0.082 \, \text{kg m}^{-3}$ found at the same height in Example 8.2 for the same current without waves. It should, however, be noted that the bed was predicted to be rippled in the cases of both current alone and wave alone, whereas it is predicted to be flat with sheet-flow in the combined case.

2. For weaker currents and/or waves, the procedure must be modified:

   - to account for the grain size in suspension, $d_{50,s}$, being smaller than that on the bed, $d_{50,b}$
   - to account for the effect of ripples by using skin-friction bed shear-stresses when calculating C_a, but total bed shear-stresses when calculating b_{max} and b_m.

3. A comparison can be made with predictions from Van Rijn's TRANSPOR program, with inputs similar to those of Example 8.4 ($h = 10$ m, $\bar{U} = 1.0 \, \text{m s}^{-1}$, $H_s = 1.0$ m, $T_p = 6$ s, $\phi = 0°$, $d_{50} = 0.2 \times 10^{-3}$ m, $d_{90} = 0.23 \times 10^{-3}$ m, $d_{50,s} = 0.2 \times 10^{-3}$ m, $10 \, °\text{C}$, $35 \, \text{ppt}$), and with the current-related and wave-related

bed-roughness heights both set to $0.01\,\mathrm{m}$ (minimum permitted value). The program outputs predictions of suspended concentration at a number of discrete heights. Applying a logarithmic interpolation to these values results in a predicted mass concentration of $C_M = 1.95\,\mathrm{kg\,m^{-3}}$ at $z = 0.10\,\mathrm{m}$, and $C_M = 0.234\,\mathrm{kg\,m^{-3}}$ at $z = 1.0\,\mathrm{m}$. These compare with the values $C_M = 2.33$ and $0.225\,\mathrm{kg\,m^{-3}}$, respectively, predicted by the method used in Example 8.4.

Bedload transport

9

9. Bedload transport

9.1. GENERAL

For current or wave flows which exceed the threshold of motion, sand moves by bedload transport. This mode of transport involves rolling, sliding and hopping (saltation) of grains along the bed, in which the weight of the grains is borne by contact with other grains rather than by the upward fluid motions which support suspended sediment. The prevalence of saltation of sand in the sea is not clear (compared to wind-driven sand transport, where it is a dominant mode of transport), although some successful bedload transport formulae are based on this concept.

Bedload transport can occur

- over a flat bed at low flows
- in conjunction with ripples or larger bedforms for stronger flows
- over a flat bed for very strong flows where ripples are washed out (sheet flow).

Bedload is the dominant mode of transport for low flow rates and/or large grains. Coarse sands and gravels are transported primarily as bedload. For stronger flows which exceed the threshold of suspension, bedload transport still occurs, but the quantity of sand carried in suspension will often be very much greater than that carried as bedload, especially for fine sands.

Bedload transport rates may be expressed in various units:

q_b = volumetric transport rate

= volume (m^3) of grains moving per unit time(s) per unit width of bed (m). SI units of q_b are thus $m^2\ s^{-1}$.

$Q_b = \rho_s q_b$ = mass transport rate $(\text{kg m}^{-1}\text{s}^{-1})$

$i_b = (\rho_s - \rho)g q_b$ = immersed weight transport rate $(\text{N m}^{-1}\text{s}^{-1})$

$q_B = q_b/(1 - \varepsilon)$ = volume of settled-bed material (including pore-water) per unit time per unit width $(\text{m}^2\text{s}^{-1})$.

Conversions to engineering units of tonnes and years for quartz material $(\rho_s = 2650\,\text{kg m}^{-3})$ and porosity $(\varepsilon = 0.40)$ can be made using the following conversion factors:

Multiply q_b in m^2s^{-1} by 3.15×10^7 to obtain q_b in $\text{m}^3\,\text{m}^{-1}\,\text{year}^{-1}$ (excluding pore water).

Multiply q_b in m^2s^{-1} by 5.26×10^7 to obtain q_B in $\text{m}^3\,\text{m}^{-1}\,\text{year}^{-1}$ (including pore water).

Multiply q_b in m^2s^{-1} by 8.36×10^7 to obtain Q_b in tonne $\text{m}^{-1}\,\text{year}^{-1}$ (excluding pore water).

9.2. BEDLOAD TRANSPORT BY CURRENTS

Knowledge

A number of bedload transport formulae have been proposed. Many of these can be expressed in the form

$$\Phi = \text{func}(\theta, \theta_{cr}) \qquad (116)$$

where

$$\Phi = \frac{q_b}{[g(s-1)d^3]^{1/2}}$$

= dimensionless bedload transport rate

$$\theta = \frac{\tau_0}{g\rho(s-1)d}$$

= Shields parameter

θ_{cr} = value of θ at threshold of motion (Section 6.4)

q_b = volumetric bedload transport rate per unit width

g = acceleration due to gravity

ρ = water density

s = ratio of densities of sediment and water

d = grain diameter.

When bedforms are present, the skin-friction component, θ_s, of θ is usually used (Section 3.3), except in the formulae of Bagnold and Van Rijn.

Some of the more commonly used or recent bedload formulae for steady flows, developed for use in rivers, are as follows:

Meyer-Peter and Müller (1948)

$$\Phi = 8(\theta - \theta_{cr})^{3/2} \qquad\qquad\text{SC (117)}$$

with $\theta_{cr} = 0\cdot047$

Bagnold (1963)

$$\Phi = F_B \theta^{1/2}(\theta - \theta_{cr}) \qquad\qquad\text{SC (118)}$$

with

$$F_B = \frac{0\cdot1}{C_D^{1/2}(\tan\phi_i + \tan\beta)}$$

$\theta = total$ Shields parameter

$C_D = total$ drag coefficient (Section 3.4)

ϕ_i = angle of repose (Section 2.2)

β = angle of slope of bed – positive for upslope flow; negative for downslope flow

Van Rijn (1984)

$$\Phi = F_R \theta^{1/2}(\theta^{1/2} - \theta_{cr}^{1/2})^{2\cdot4} \qquad\qquad (119)$$

with

$$F_R = \frac{0\cdot005}{C_D^{1\cdot7}}\left(\frac{d}{h}\right)^{0\cdot2}$$

and C_D and θ defined as in Bagnold's formula

Yalin (1963)

$$\Phi = F_Y \theta^{1/2}(\theta - \theta_{cr}) \qquad\qquad\text{SC (120)}$$

with

$$F_Y = \frac{0\cdot635}{\theta_{cr}}\left[1 - \frac{1}{aT}\ln(1 + aT)\right]$$

$$a = 2\cdot45\theta_{cr}^{0\cdot5}\,s^{-0\cdot4}$$

$$T = (\theta - \theta_{cr})/\theta_{cr}$$

Madsen (1991)

$$\Phi = F_M(\theta^{1/2} - 0\cdot7\theta_{cr}^{1/2})(\theta - \theta_{cr}) \tag{121}$$

with

$F_M = 8/\tan \phi_i$ for rolling/sliding grains

$F_M = 9\cdot5$ for saltating sand grains in water

Ashida and Michiue (1972)

$$\Phi = 17(\theta^{1/2} - \theta_{cr}^{1/2})(\theta - \theta_{cr}) \tag{122}$$

Wilson (1966)

$$\Phi = 12\theta^{3/2} \tag{123}$$

Nielsen (1992)

$$\Phi = 12\theta^{1/2}(\theta - \theta_{cr}) \qquad\qquad \text{SC} \tag{124}$$

In all the Equations (117)–(124) except (123) it is understood that $\Phi = 0$ if $\theta \leq \theta_{cr}$ (i.e. no transport if the flow is below the threshold of motion). Equation (123) is intended only for sheet-flow conditions, for which $\theta \gg \theta_{cr}$.

Equation (124) was proposed by Nielsen (1992) by fitting to bedload transport data, and was independently derived by Soulsby on the basis of sheet flow theory and experiments.

There are clear similarities between Equations (117)–(124), varying mainly in the form of the initial coefficient. Figure 29 shows that Equation (124) gives a good fit to data (with $\theta_{cr} = 0\cdot05$, typical of the sediments plotted).

Although the above formulae were developed for steady flows in rivers, they can be used on an instantaneous basis in tidal flows, and under waves or combined waves and currents, because the response time of a sand grain in bedload motion is very short compared to a tidal period or a wave period.

Further discussion of the bedload formulae of Meyer-Peter and Müller, Bagnold, Einstein and Yalin is given by Fisher (1993), in the context of sediment transport in rivers.

For a sloping bed, the Bagnold formula can be used. Parker and Kovacs (1993) have also modified the Ashida and Michiue formula for beds of general longitudinal and lateral slope, and give a computer program on diskette for its implementation.

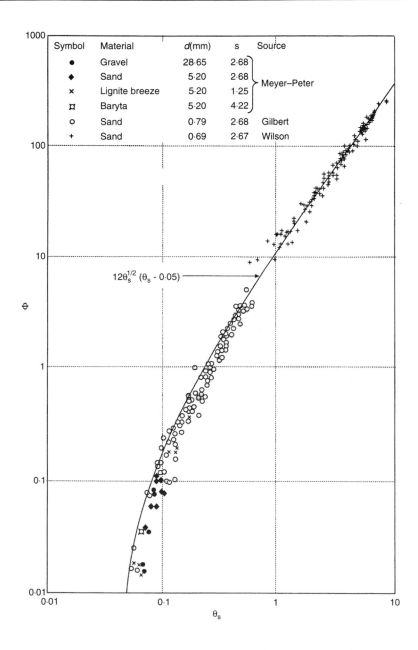

Figure 29. Bedload transport rate of sand/gravel for steady current (modified from Nielsen (1992))

Procedure

1. *Example 9.1. Bedload transport by currents*

   Obtain estimates of the
   following parameters:

depth-averaged current speed	$\bar{U}$	$1 \cdot 0 \text{ m s}^{-1}$
water depth	h	5 m
grain diameter (take $d = d_{50}$)	d	1 mm
density ratio of sediment and water	s	$2 \cdot 58$
kinematic viscosity	ν	$1 \cdot 36 \times 10^{-6} \text{ m}^2 \text{ s}^{-1}$

   Calculate bed shear-stress
   for flat bed (see Example 3.3) τ_0 $1 \cdot 84 \text{ N m}^{-2}$

   or   if bedforms are present,
           calculate skin-friction
           component τ_{0s} (see
           Section 3.3).

   Calculate Shields parameter

   $$\theta = \frac{\tau_0}{g\rho(s-1)d}$$ $0 \cdot 115$

   Calculate dimensionless
   grain size

   $$D_* = d \left[\frac{g(s-1)}{\nu^2} \right]^{1/3}$$ $20 \cdot 3$

   and hence θ_{cr} from Equation (77) $0 \cdot 0302$

   For Nielsen (1992) method:

   Calculate Φ from Equation (124) $0 \cdot 348$

   Calculate $q_b = \Phi[g(s-1)d^3]^{1/2}$ $43 \cdot 3 \times 10^{-6} \text{ m}^2 \text{ s}^{-1}$

2. For comparison, values of q_b from Equations (117) to (124) for
   these inputs are

Meyer-Peter and Müller	q_b	17.8×10^{-6} m^2 s^{-1}
Bagnold ($\phi_i = 32°, \beta = 0°$)	q_b	10.9×10^{-6} m^2 s^{-1}
Van Rijn	q_b	23.9×10^{-6} m^2 s^{-1}
Yalin	q_b	20.5×10^{-6} m^2 s^{-1}
Madsen – rolling ($\phi_i = 32°$)	q_b	29.4×10^{-6} m^2 s^{-1}
– saltating	q_b	21.8×10^{-6} m^2 s^{-1}
Ashida and Michiue	q_b	29.7×10^{-6} m^2 s^{-1}
Wilson	q_b	46.8×10^{-6} m^2 s^{-1}
Nielsen	q_b	43.3×10^{-6} m^2 s^{-1}

There is a variation of a factor of more than 4 between the largest and the smallest of these values, illustrating the degree of uncertainty in predictions of bedload transport rate.

9.3.   BEDLOAD TRANSPORT BY WAVES

Knowledge

Under *waves*, the net bedload transport is zero if the oscillatory velocity is symmetrical (e.g. sinusoidal). However, the transport of sediment during half of a wave cycle is of interest in some applications. In asymmetrical wave motions, such as occur under steep waves in shallow water (see Section 4.4), there is a non-zero net sediment transport because transport beneath the crest is greater than that beneath the trough, resulting in a net sediment transport in the direction of wave travel. The net transport can then be calculated as the difference between the half-cycle transport beneath the crest and the half-cycle transport beneath the trough.

The following formulae for the wave half-cycle volumetric sediment transport rate $q_{b1/2}$ have been proposed:

Madsen and Grant (1976): $q_{b1/2} = F_{MG} w_s d\theta_w^3$ \hfill (125)

with \hfill $F_{MG} = 12.5$ for $\theta_w \gg \theta_{cr}$,

and \hfill $F_{MG} \rightarrow 0$ as $\quad \theta_w \rightarrow \theta_{cr}$

Sleath (1978):

(for coarse grains) \hfill $q_{b1/2} = 47\omega d^2(\theta_w - \theta_{cr})^{3/2}$ \quad SC (126)

Soulsby: \hfill $q_{b1/2} = 5.1[g(s-1)d^3]^{1/2}(\theta_w - \theta_{cr})^{3/2}$

\hfill SC (127)

where

θ_w = amplitude of oscillatory component of θ due to wave (use skin-friction component if bed is rippled)

w_s = settling velocity of grain (Section 8.3)

ω = radian wave frequency

θ_{cr}, g, ρ, s and d as defined in Section 9.2.

As for steady flows, $q_{b1/2} = 0$ if $\theta_w \leq \theta_{cr}$ in Equations (126) and (127).

Equation (127) is obtained by integrating Equation (124) through half a wave cycle, and is similar to a formula given by Sleath (1982) for fine grains (Sleath's coefficient was 1·95 instead of 5·1).

Procedure

1. *Example 9.2. Bedload transport by waves*

To calculate the net bedload transport by waves, obtain values of:

water depth	h	5 m
median grain diameter	d_{50}	1 mm
wave height	H	1 m
wave period	T	6 s
Calculate orbital velocity amplitude using linear wave theory (see Example 4.2)	U_w	0·569 m s^{-1}
Calculate wavelength (see Example 4.1)	L	38·1 m
Calculate $2\pi h/L =$	kh	0·825
Use Stokes' second-order wave theory to calculate orbital velocity under wave-crest (Equation (55a))	U_{wc}	0·614 m s^{-1}

and wave-trough (Equation (55b)) U_{wt}		$0 \cdot 524$ m s^{-1}

Calculate the skin-friction bed shear-stress (e.g. use the Soulsby method; see Example 4.4)

under the wave-crest τ_{ws}(crest)		$2 \cdot 69$ N m^{-2}
under the wave-trough τ_{ws}(trough)		$2 \cdot 13$ N m^{-2}

Convert to skin-friction Shields parameter, using Equation (2a) with $\rho = 1027$ kg m^{-3}, $\rho_s = 2650$ kg m^{-3}

under crest	θ_{ws}(crest)	$0 \cdot 169$
under trough	θ_{ws}(trough)	$0 \cdot 134$

Calculate threshold bed shear-stress (e.g. use the Soulsby method; see Example 6.3) θ_{cr} — $0 \cdot 0302$

Calculate the wave half-cycle sediment transport rate, e.g. use the Soulsby formula (Equation (127))

under crest	$q_{b1/2}$ (crest)	$32 \cdot 8 \times 10^{-6}$ m^2 s^{-1}
under trough	$q_{b1/2}$ (trough)	$21 \cdot 1 \times 10^{-6}$ m^2 s^{-1}

Calculate the net transport rate in the direction of wave travel $= q_{b1/2}$ (crest) $- q_{b1/2}$(trough) q_b — $11 \cdot 7 \times 10^{-6}$ m^2 s^{-1}

2. For comparison, calculations with the same inputs to the Madsen and Grant formula (Equation (125), with $F_M = 12 \cdot 5$) yields

$q_b = 3\cdot47 \times 10^{-6}$ m^2 s^{-1}, and to the Sleath formula (Equation (126)) yields $q_b = 0\cdot905 \times 10^{-6}$ m^2 s^{-1}. Thus there is a variation of more than a factor 10 between methods in this case.

3. Note that the effective period used for both the crest and trough half-periods is T; a more detailed analysis using a longer period under the trough and a shorter period under the crest would be more correct, but is not really justified in view of the other uncertainties in the calculations.

4. The above procedure takes account of only wave asymmetry as a mechanism for net sediment transport. Other mechanisms which might be present, and could cause net transport of similar or greater magnitude, include superimposed currents (tidal, wind-driven, longshore), undertow and mass transport (streaming). See Section 10.3 for further details.

9.4. BEDLOAD TRANSPORT BY WAVES PLUS CURRENTS

Knowledge

For *combined waves and currents*, the waves provide a stirring mechanism which keeps the sediment grains mobile, while the current adds to the stirring and also provides a mechanism for net transport. The non-linear interaction of the bed shear-stresses (see Section 5.3) is an important factor in determining the bedload transport. The following formulae give the net bedload transport rate averaged over a sinusoidal wave-cycle.

Bijker (1967)

$$q_b = A_B u_* d \exp\left[\frac{-0\cdot27 g(s-1)d}{\mu(u_*^2 + 0\cdot016 U_w^2)}\right] \qquad \text{SC (128)}$$

with

$u_* = $ (total) friction velocity due to current alone

$$= \frac{0\cdot40\bar{U}}{\ln(12h/\Delta_r)}$$

$\mu = $ 'ripple factor'

$$= \left[\frac{\ln(12h/\Delta_r)}{\ln(12h/d_{90})}\right]^{1\cdot5}$$

U_w = wave orbital velocity amplitude

A_B = 1 for non-breaking waves; 5 for breaking waves.

This is the earliest sediment transport formula for combined waves and currents, and is still widely used. It was derived by determining the bed shear-stress through a wave–current interaction model (see Section 5.3), and using this to modify an existing sediment transport formula for current alone by Kalinske and Frijlink.

Similar formulae were derived by other researchers (Swart, Willis, Van de Graaf and Van Overeem), using the Bijker wave-plus-current shear-stress to modify the current-alone formulae of Engelund and Hansen (1972), Ackers and White (1973), etc. A test by Bettess (1985) of these formulae against laboratory and field data showed that none of them gave particularly good agreement.

A more recent formula is that of Soulsby:

$$\Phi_{x1} = 12\theta_m^{1/2}(\theta_m - \theta_{cr}) \qquad\qquad \text{SC} \,(129a)$$

$$\Phi_{x2} = 12(0{\cdot}95 + 0{\cdot}19 \cos 2\phi)\theta_w^{1/2}\theta_m \qquad\qquad \text{SC} \,(129b)$$

$$\Phi_x = \text{maximum of } \Phi_{x1} \text{ and } \Phi_{x2} \qquad\qquad \text{SC} \,(129c)$$

$$\Phi_y = \frac{12(0{\cdot}19\theta_m\theta_w^2 \sin 2\phi)}{\theta_w^{3/2} + 1{\cdot}5\theta_m^{3/2}} \qquad\qquad \text{SC} \,(129d)$$

subject to $\Phi_x = \Phi_y = 0$ if $\theta_{max} \leq \theta_{cr}$

where

$$\Phi_{x,y} = \frac{q_{bx,y}}{[g(s-1)d^3]^{1/2}}$$

q_b = mean volumetric bedload transport rate per unit width

q_{bx} = component of q_b travelling in the direction of the current

q_{by} = component of q_b travelling at right angles to the current, in the same sense as the angle ϕ

θ_m = mean value of θ over a wave cycle

θ_w = amplitude of oscillatory component of θ due to wave

ϕ = angle between current direction and direction of wave travel

$$\theta_{max} = [(\theta_m + \theta_w \cos \phi)^2 + (\theta_w \sin \phi)^2]^{1/2}$$

Use skin-friction components of θ_w, θ_m and θ_{max} if bed is rippled.

Equations (129a–d) are an approximation to the bedload transport obtained by integrating Equation (124) through one wave cycle, in which the oscillatory component of θ has a sinusoidal time dependence.

The effect of the current enters through θ_m, and of the waves through θ_w. Equation (129a) corresponds to current-dominated conditions, and Equation (129b) to wave-dominated conditions. The value of θ_m is obtained by calculating τ_m by one of the wave-current interaction methods given in Section 5.3.

The term Φ_y is non-zero for $\phi \neq 0°$, 90°, 180° or 270°, showing that a current with oblique waves causes a transverse component of sediment transport; for oblique waves driving a longshore current, this component of transport is directed onshore.

Soulsby–Van Rijn

The bedload part of the total-load formula given in Section 10.4 (*Knowledge*) can be used on its own, by using just the term A_{sb} (Equation (136b)) in Equation (136a).

Procedure

1. To calculate the mean bedload transport due to combined waves and currents it is necessary to obtain input values of the same quantities required for current alone (Section 9.2), supplemented by those required for waves alone (Section 9.3). The procedure is illustrated using the bedload transport formula of Soulsby with input values taken from Example 9.1 (current) and Example 9.2 (waves). The waves are taken to be directed at 45° to the current, over a flat bed of 1 mm quartz sand in sea water at 10 °C and 35 ppt. The transport due to wave asymmetry, illustrated in Example 9.2, is neglected in the present example.

2. *Example 9.3. Bedload transport by waves and currents*

Obtain values of the
following parameters:

water depth	h	5 m
median grain diameter	d_{50}	1 mm
density ratio of sediment and water	s	2·58
kinematic viscosity	ν	$1·36 \times 10^{-6}\,\mathrm{m^2\,s^{-1}}$
depth-averaged current speed	$\bar{U}$	$1·0\,\mathrm{m\,s^{-1}}$
wave height	H	1 m
wave period	T	6 s
angle between wave and current directions	ϕ	45°
Calculate orbital velocity (Example 9.2)	U_{w}	$0·569\,\mathrm{m\,s^{-1}}$
Calculate the skin-friction wave shear-stress (e.g. use the Soulsby method)	τ_{ws}	$2·40\,\mathrm{N\,m^{-2}}$
and Shields parameter	θ_{ws}	0·151
Calculate the mean skin-friction shear-stress due to combined waves and current (e.g. use DATA13 method; see Example 5.1)	τ_{ms}	$1·899\,\mathrm{N\,m^{-2}}$
and Shields parameter	θ_{ms}	0·119
Calculate threshold Shields parameter (e.g. use Soulsby method, see Example 6.3)	θ_{cr}	0·0302
Calculate mean bedload transport rate using the Soulsby method (Equations (129a–d))		
Equation (129a)	Φ_{x1}	0·368

Equation (129*b*)	Φ_{x2}	0·527
Equation (129*c*) (max. of Φ_{x1} and Φ_{x2})	Φ_x	0·527
Equation (129*d*)	Φ_y	0·0514
Calculate $[g(s-1)d^3]^{1/2}$		$1\cdot25 \times 10^{-4}\,\mathrm{m^2\,s^{-1}}$
Multiply by Φ_x and Φ_y to obtain:		
bedload transport rate in direction of current	q_{bx}	$65\cdot7 \times 10^{-6}\,\mathrm{m^2\,s^{-1}}$
bedload transport rate at right angles to current	q_{by}	$6\cdot39 \times 10^{-6}\,\mathrm{m^2\,s^{-1}}$

3. Comparing this example with Example 9.1, which gave $q_b = 43\cdot3 \times 10^{-6}\,\mathrm{m^2\,s^{-1}}$ for the same current without waves, the addition of the waves has increased the mean transport rate in the current direction by about 50%. This increase is about twice as large as the transport due to wave asymmetry ($q_b = 11\cdot7 \times 10^{-6}\,\mathrm{m^2\,s^{-1}}$) calculated in Example 9.2. In addition, there is a transport at right angles to the current which has a magnitude of about 10% of that in the current direction. If the current was a longshore current driven by the waves incident at 45° to a coastline, then q_{by} would give an onshore component of transport.

4. For comparison, calculations with the same inputs plus $d_{90} = 1\cdot56\,\mathrm{mm}$ to the Bijker formula (Equation (128)), with $A_B = 1$) yields $q_b = 20\cdot2 \times 10^{-6}\,\mathrm{m^2\,s^{-1}}$, and to the bedload part of the Soulsby–Van Rijn formula (Equation (136)) yields $q_b = 179 \times 10^{-6}\,\mathrm{m^2\,s^{-1}}$. There is thus a difference of a factor of 9 between the largest and smallest estimates in this example, with the Soulsby formula (Equation (129)) lying in the middle.

Total load transport

10

10. Total load transport

10.1. GENERAL

The total sediment transport rate is the quantity required most commonly for addressing practical applications such as infill of dredged channels, dispersion of spoil-heaps, and morphodynamic response of coastal areas to engineering works.

The bedload and suspended load contributions to the total sediment transport rate may be calculated separately and added (in which case the two contributions must be compatible and matched at a well-defined height). Alternatively the total sediment transport rate may be given by an empirical formula.

The general principle for the calculation of the suspended sediment transport rate q_s is to make an integral through the water depth of the sediment flux $U(z)C(z)$ at height z, having first established the vertical profiles of velocity $U(z)$ (see Section 3.2) and concentration $C(z)$ (see Chapter 8). Thus

$$q_s = \int_{z_a}^{h} U(z)C(z) \, dz \tag{130}$$

where $z_a =$ reference height near the bed, at which the reference concentration C_a is calculated, and corresponding to the top of the bedload layer

$h =$ water depth

For combined waves and currents, $U(z)$ and $C(z)$ are usually taken to be the mean values over a wave cycle, although strictly there is also a contribution due to the covariance of the time-varying velocities and concentrations. The covariance contribution can be quite large, and is often in the opposite direction to the current, so that the net sediment transport is reduced or even negative. It is greatest for the case of large waves with a small current. However, means of calculating this contribution are not yet well-established, and it will not be considered in this book.

The accuracy of sediment transport formulae is not high (see Section 1.6). In rivers, the best formulae give predictions which are within a factor of 2 of the observed value in no better than about 70% of samples. In the sea the position is worse— probably no better than a factor of 5 in 70% of samples. The accuracy can be improved significantly by making site specific measurements and calibrating the formulae to match.

Total load transport rates may be expressed in various units:

q_t = volumetric transport rate
= volume (m^3) of grains moving per unit time (s) per unit width of bed (m); SI units of q_t are thus m^2 s^{-1}

$Q_t = \rho_s q_t$ = mass transport rate (kg m^{-1} s^{-1})

$i_t = (\rho_s - \rho)g q_t$ = immersed weight transport rate (N m^{-1} s^{-1})

$q_T = q_t/(1 - \varepsilon)$ = volume of settled-bed material (including pore water) per unit time per unit width (m^2 s^{-1}).

See Section 9.1 for conversions to engineering units.

10.2. TOTAL LOAD TRANSPORT BY CURRENTS

Knowledge

For tidal currents in the absence of waves (e.g. in enclosed estuaries) it is common practice to use sediment transport formulae developed for rivers, in a quasi-steady fashion. There are over 20 of these (see Sleath, 1984, for a fuller list), of which the three quoted below are the most widely used and best suited for the sea. They are all less appropriate for use in the sea than in rivers, because they assume that the sandwaves/dunes are in equilibrium with the instantaneous flow velocity, which is not true in tidal flows.

Engelund and Hansen (1972)

$$q_t = \frac{0 \cdot 04 C_D^{3/2} \bar{U}^5}{[g(s - 1)]^2 d_{50}} \qquad \text{SC (131)}$$

where the drag coefficient C_D must be determined by the Engelund (1966) alluvial friction method. This method was originally derived by considering the energy balance for bedload transport over dunes, but is now widely used (and reasonably accurate) as a total load equation. It does not include a threshold of motion condition.

Ackers and White (1973)

$$q_t = C_{AW}\bar{U}d\left(\frac{\bar{U}}{u_*}\right)^n \left(\frac{F_{AW} - A_{AW}}{A_{AW}}\right)^m \qquad \text{SC } (132a)$$

where

$$F_{AW} = \frac{u_*^n}{[g(s-1)d]^{1/2}} \left[\frac{\bar{U}}{2\cdot46 \ln(10h/d)}\right]^{1-n} \qquad \text{SC } (132b)$$

and if

$$D_* = \left[\frac{g(s-1)}{\nu^2}\right]^{1/3} d \qquad \text{SC } (132c)$$

then, for $1 < D_* \leq 60$ (fine sediments)

$$n = 1 - 0\cdot243 \ln D_*$$

$$A_{AW} = \frac{0\cdot23}{D_*^{1/2}} + 0\cdot14$$

$$m = \frac{9\cdot66}{D_*} + 1\cdot34$$

$$C_{AW} = \exp\left[2\cdot86 \ln D_* - 0\cdot434 (\ln D_*)^2 - 8\cdot13\right] \qquad (132d)$$

and, for $D_* > 60$ (coarse sediments),

$$n = 0; \quad A_{AW} = 0\cdot17; \quad m = 1\cdot5; \quad C_{AW} = 0\cdot025 \qquad (132e)$$

The grain size d must be taken as $d = d_{35}$. The friction velocity $u_* = C_D^{1/2}\bar{U}$ must be determined by the White et al. (1980) alluvial friction method. The transport formula was derived by considering the form of the transport relations for bedload (coarse sediments) and suspended load (fine sediments) separately, and uniting them through a transition in the range $1 < D_* \leq 60$ through the empirical coefficients n, A_{AW}, m and C_{AW}, which were fitted to a large data set.

A revised set of values of the coefficients based on more recent data was derived in 1990. The expressions for n and A_{AW} are unchanged, but the revised expressions for m and C_{AW} are:

for $\quad 1 \le D_* \le 60$

$$m = \frac{6\cdot83}{D_*} + 1\cdot67$$

$$C_{AW} = \exp\left[2\cdot79 \ln D_* - 0\cdot426(\ln D_*)^2 - 7\cdot97\right] \quad \text{SC}(132f)$$

and for $\quad D_* > 60$

$$m = 1\cdot78; \quad C_{AW} = 0\cdot025 \qquad\qquad \text{SC}(132g)$$

Van Rijn (1984)

Van Rijn derived a full, comprehensive theory of sediment transport in rivers based on a mixture of fundamental physics and empirical results. He parameterised the results of the full method (to $\pm25\%$) in the following simpler formulae:

$$q_t = q_b + q_s \qquad\qquad\qquad \text{SC}(133a)$$

$$q_b = 0\cdot005\bar{U}h\left\{\frac{\bar{U} - \bar{U}_{cr}}{[(s-1)gd_{50}]^{1/2}}\right\}^{2\cdot4}\left(\frac{d_{50}}{h}\right)^{1\cdot2} \qquad \text{SC}(133b)$$

$$q_s = 0\cdot012\bar{U}h\left\{\frac{\bar{U} - \bar{U}_{cr}}{[(s-1)gd_{50}]^{1/2}}\right\}^{2\cdot4}\left(\frac{d_{50}}{h}\right)(D_*)^{-0\cdot6} \quad \text{SC}(133c)$$

with

$$\bar{U}_{cr} = 0\cdot19(d_{50})^{0\cdot1}\log_{10}\left(\frac{4h}{d_{90}}\right) \quad \text{for } 0\cdot1 \le d_{50} \le 0\cdot5 \text{ mm}$$
$$(133d)$$

$$\bar{U}_{cr} = 8\cdot5(d_{50})^{0\cdot6}\log_{10}\left(\frac{4h}{d_{90}}\right) \quad \text{for } 0\cdot5 \le d_{50} \le 2 \text{ mm} \quad (133e)$$

SI units must be used in Equations (133d) and (133e).

The formulae are valid for parameters in the ranges $h = 1{-}20$ m, $\bar{U} = 0\cdot5$ to 5 m s^{-1}, $d_{50} = 0\cdot1$ to 2 mm and for fresh water at $15\,^\circ$C.

Equation (133b) is the same as Equation (119) for bedload transport.

This is the form most easily used in the sea. The full Van Rijn (1984) method is given in step-by-step form by Van Rijn (1993, Appendix A), or by Fisher (1993), and involves much more computation, including the calculation of the alluvial friction by the Van Rijn (1984) method.

All three methods were originally derived for application to rivers, and are less readily applied in the sea because the water surface slope is not usually an available input parameter, nor is it uniquely related to the bed friction. The bed shear-stress must be calculated by other means in the sea.

A comparison of the methods of Engelund and Hansen, Ackers and White (original) and Van Rijn (full method) is shown in Figure 30, for the specific case of quartz sand with the grain-size distribution given in Example 10.1 in fresh 20 °C water of depth 10 m for current speeds up to $1.4\,\mathrm{m\,s^{-1}}$. The bed friction was predicted by the methods of Engelund; White, Paris and Bettess; and Van Rijn, respectively. The calculations were made using the SandCalc software package. All three methods show very small transport rates for $U < 0.5\,\mathrm{m\,s^{-1}}$, and a rapid increase in sediment transport with current speed for currents in excess of $0.5\,\mathrm{m\,s^{-1}}$. At $1.4\,\mathrm{m\,s^{-1}}$, the Ackers and White method predicts approximately twice, and the Engelund and Hansen

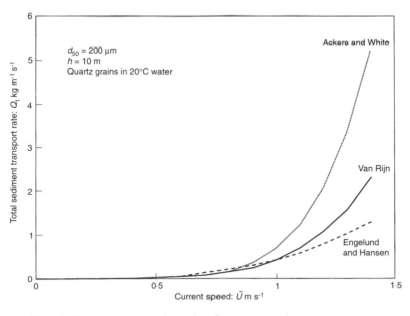

Figure 30. Sediment transport formulae for currents alone

method approximately half, the value given by the Van Rijn method. There are many other methods, but these three perform about equally well when compared with large data sets, with 60% to 70% of predictions lying within a factor of 2 of the observed values. It is unlikely that this degree of accuracy will be improved upon significantly, since intercomparison of the data itself shows about this degree of variability for fixed values of the input variables.

Procedure

1. *Example 10.1. Total load transport by currents*

To calculate the total sediment transport rate q_t (volume/volume) for a current alone, obtain values of

bed sediment size grading (different formulae require different percentiles)	d_{16}	141 μm
	d_{35}	175 μm
	d_{50}	200 μm
	d_{65}	230 μm
	d_{84}	282 μm
	d_{90}	315 μm
water depth	h	10 m
depth-averaged current velocity	$\bar{U}$	$1{\cdot}0 \text{ m s}^{-1}$
sediment density	ρ_s	2650 kg m^{-3}
temperature		20 °C
salinity		0 ppt
acceleration due to gravity	g	$9{\cdot}81 \text{ m s}^{-2}$
Calculate kinematic viscosity (Section 2.1)	ν	$1{\cdot}0 \times 10^{-6} \text{ m}^2 \text{ s}^{-1}$
Calculate water density (Section 2.1)	ρ	1000 kg m^{-3}
Calculate sediment density ratio s		$2{\cdot}65$

2. Detailed worked examples of the Ackers and White, and the Van Rijn, methods are given in *Manual of Sediment Transport in Rivers* (Fisher, 1993). The numerical values given in Example 10.1 lead to the following sediment transport rates predicted by the various formulae, as calculated using SandCalc:

Engelund and Hansen $\qquad q_t = 1 \cdot 52 \times 10^{-4} \, \mathrm{m^2 \, s^{-1}}$

Ackers and White (original) $\qquad q_t = 2 \cdot 53 \times 10^{-4} \, \mathrm{m^2 \, s^{-1}}$

Ackers and White (revised) $\qquad q_t = 1 \cdot 98 \times 10^{-4} \, \mathrm{m^2 \, s^{-1}}$

Van Rijn (full method) $\qquad q_t = 1 \cdot 40 \times 10^{-4} \, \mathrm{m^2 \, s^{-1}}$

Van Rijn (formula) $\qquad q_t = 2 \cdot 75 \times 10^{-4} \, \mathrm{m^2 \, s^{-1}}$

There is a factor of 2 between the smallest and largest predicted values, illustrating the degree of uncertainty in calculating total sediment transport rates in steady flows.

10.3. TOTAL LOAD TRANSPORT BY WAVES

Knowledge

Waves cause a net sediment transport by a number of mechanisms:

- Waves entrain sediment more readily than currents, and diffuse it through the wave boundary layer. In the presence of a current the current-generated turbulence diffuses the suspended sediment to higher levels, and carries it with the net flow.
- Waves generate longshore currents when they break on a beach, which transport the sediment (by the above mechanism) along the coast.
- The orbital velocity beneath a wave crest is larger than that beneath a wave trough, and hence entrains more sediment. This causes a net sediment transport in the direction of wave travel (onshore, generally).
- A mass-transport (or streaming) of water in the direction of wave-travel is produced within the wave boundary layer, which carries sediment in the wave direction.

- Waves produce an offshore undertow velocity near the bed within the surf zone, which carries sediment offshore.

One of the most widely used methods is the Bailard (1981) formula, in which the total sediment transport vector q_t is made up of the sum of four terms:

$$q_t = q_{bo} - q_{bs} + q_{so} - q_{ss} \tag{134a}$$

with

$$q_{bo} = \frac{c_f \varepsilon_B}{g(s-1)\tan \phi_i} \langle |u|^2 u \rangle \tag{134b}$$

$$q_{bs} = \frac{c_f \varepsilon_B \tan \beta}{g(s-1)\tan^2 \phi_i} \langle |u|^3 \rangle i \tag{134c}$$

$$q_{so} = \frac{c_f \varepsilon_s}{g(s-1)w_s} \langle |u|^3 u \rangle \tag{134d}$$

$$q_{ss} = \frac{c_f \varepsilon_s^2 \tan \beta}{g(s-1)w_s^2} \langle |u|^5 \rangle i \tag{134e}$$

where

g = acceleration due to gravity
ρ_s = sediment density
$s = \rho_s/\rho$
ρ = water density
c_f = friction coefficient such that $\tau = \rho c_f |u| u$
τ = bed shear-stress vector
u = total near-bottom velocity due to combined waves and currents
ϕ_i = angle of internal friction of sediment ($\tan \phi_i = 0.63$)
$\tan \beta$ = bed slope
i = unit vector directed upslope
w_s = sediment settling velocity
ε_B = efficiency of bedload transport $= 0.10$
ε_s = efficiency of suspended transport $= 0.02$
$\langle . \rangle$ = is a time average over many waves.

The terms represent: q_{bo} = bedload transport on horizontal bed, q_{bs} = slope effect on bedload transport, q_{so} = suspended load on horizontal bed, q_{ss} = slope effect on suspended load.

The Bailard formula was developed from the 'energetics' arguments proposed successively by Bagnold, Inman and Bowen; the general approach is that the work done in transporting sediment is a fixed proportion of the total energy dissipated by the flow. It was originally devised for cross-shore transport and longshore transport in the surf zone. The formula applies to the transport at a point: to obtain the total longshore transport rate, for example, the sediment transport rate must be integrated across the surf zone.

It is popular with numerical modellers because it is computationally efficient, and it takes account of

- bedload and suspended load transport
- waves and currents, including the effects of wave asymmetry
- bed slopes in any direction.

Two aspects of implementation are not specified in the original formula:

- the height at which the total velocity u is measured; a height of 0·05 m seems to be suitable
- the form of the friction coefficient c_f; Bailard derived this from data on a case-by-case basis, but other possibilities are to use $0·5 f_w$ (see Section 4.5), or better still a form related to combined wave-plus-current bed shear-stress (see Section 5.3).

Soulsby (1995b) reported the results of an extensive series of tests of the Bailard formula against laboratory and field data and against more sophisticated sediment transport models, made by a task force of researchers from six European organisations. They were investigating the validity of the method for use in coastal morphodynamic models outside as well as inside the surf zone. Their principal findings were as follows:

- There is a wide range of interpretation of the definition of the Bailard formula, and previously reported good agreements with data have often relied on individual tuning.
- For the flat-bed, surf-zone conditions for which the Bailard formula was originally devised, predicted transport was generally within a factor of 5, and often within a factor of 2,

of the observed values. The friction coefficient must be based on a grain-related roughness in this case.

- Over rippled beds the predictions can be up to a factor of 100 in error, even if a ripple-related roughness is used. Worse still, both laboratory and field measurements often showed transport in the opposite direction to that predicted. Better results may be obtained by taking velocities within 1 mm to 5 mm of the bed and using a grain-related roughness, although this may be impractical in morphodynamic models.
- Agreement was better for wave-dominated than for current-dominated conditions.
- Comparisons with other mathematical models showed reasonably good agreement, suggesting that the more sophisticated models may suffer similar deficiencies to the Bailard formula in their ability to make accurate predictions over a wide range of conditions.

10.4. TOTAL LOAD TRANSPORT BY WAVES PLUS CURRENTS

Knowledge

The majority of sand transport calculations in coastal and offshore areas involve both waves and currents as driving forces. In essence, the waves (with some help from the current) stir up the sand, and the current transports it. Waves may themselves generate currents, either as longshore currents through the release of radiation stresses during the breaking process on a beach, or as mass transport (or streaming) in the wave direction of travel near the bed, as a result of boundary-layer processes.

The Bailard formula (Section 10.3) can be used for combined waves and currents, as well as for asymmetrical waves. Other methods specifically derived for combined waves and currents are given below.

Grass (1981)
If the sediment transport rate q_{tc} due to a current alone is given by

$$q_{tc} = A_G \bar{U}^n \tag{135a}$$

then the transport rate q_t (volume/volume) due to combined waves and current is given by

$$q_t = A_G \bar{U} \left(\bar{U}^2 + \frac{0.08}{C_D} U_{rms}^2 \right)^{(n-1)/2}$$

(135b)

where

$A_G, n =$ empirical coefficients obtained by fitting Equation (135a) to site-specific data

$\bar{U} =$ depth-averaged current speed

$U_{rms} =$ root-mean-square wave orbital velocity

$C_D =$ drag coefficient due to current alone.

The formula was derived for suspended load transport by considering the turbulent kinetic energy produced by the combined wave and current boundary layers. It is particularly useful where site-specific measurements of suspended sediment transport rate have been made in calm conditions to obtain values of A_G and n, as an extrapolation procedure to make predictions for stormy conditions. However, it should only be used for conditions with $U_{rms} < \bar{U}$, and for rippled beds, corresponding to the field data against which the coefficient 0.08 was calibrated.

Soulsby–Van Rijn

$$q_t = A_s \bar{U} \left[\left(\bar{U}^2 + \frac{0.018}{C_D} U_{rms}^2 \right)^{1/2} - \bar{U}_{cr} \right]^{2.4} (1 - 1.6 \tan \beta)$$

SC (136a)

where

$$A_{sb} = \frac{0.005h(d_{50}/h)^{1.2}}{[(s-1)gd_{50}]^{1.2}}$$

SC (136b)

$$A_{ss} = \frac{0.012d_{50}D_*^{-0.6}}{[(s-1)gd_{50}]^{1.2}}$$

SC (136c)

$$A_s = A_{sb} + A_{ss}$$

SC (136d)

$\bar{U} =$ depth-averaged current velocity

$U_{rms} =$ root-mean-square wave orbital velocity

$$C_D = \left[\frac{0.40}{\ln(h/z_0) - 1} \right]^2$$

$=$ drag coefficient due to current alone

$\bar{U}_{cr}$ = threshold current velocity from Equations (133d,e)
β = slope of bed in streamwise direction, positive if flow runs uphill
h = water depth
d_{50} = median grain diameter
z_0 = bed roughness length = 0·006 m
s = relative density of sediment
g = acceleration due to gravity
ν = kinematic viscosity of water

$$D_* = \left[\frac{g(s-1)}{\nu^2}\right]^{1/3} d_{50}$$

The formula applies to total (bedload plus suspended load) sediment transport in combined waves and currents on horizontal and sloping beds. Term A_{sb} gives the bedload, and term A_{ss} the suspended load, transport. The method is intended for conditions in which the bed is rippled, and z_0 should be set to 6 mm.

It was derived by applying the principles used by Grass to the current-alone formula of Van Rijn (see above), and modifying to include a threshold term and a slope term. The coefficient 0·018 differs from the value 0·08 in Equation (135b), and was obtained by calibration against the curves plotted by Van Rijn (1993, Appendix A) which gave transport rates calculated using his TRANSPOR program. It has the benefits over the Grass formula of: giving a value for the coefficient A_s in the absence of site-specific measurements; including bedload and slope effects; and including a threshold velocity. A comparison of the Soulsby–Van Rijn formula and Van Rijn's TRANSPOR program is shown in Figure 31 for a range of current speeds and wave conditions. The large enhancement of the transport rate due to the wave action is evident.

The bed-slope term $(1 - 1·6 \tan \beta)$ is a device commonly used by mathematical and numerical modellers, but is a less correct procedure than modifying the threshold velocity for slope effects (see Section 6.4).

Van Rijn (1989)

Van Rijn adapted the current-alone Van Rijn formula by including an analytical, semi-empirical, model of the diffusion of sediment through the wave boundary layer. It is rather

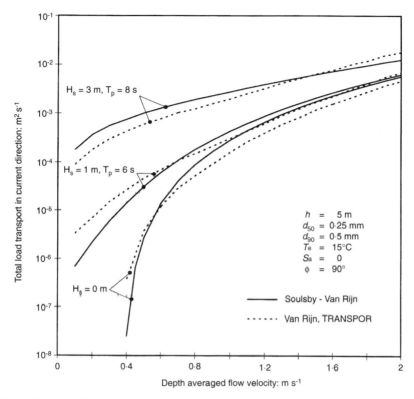

Figure 31. Sediment transport by waves plus currents

complicated to present in full here (see Van Rijn, 1993 for full
details), and is most easily applied by using the computer
program TRANSPOR on a disk supplied with Van Rijn's book.
The principles involved include the following.

- The bedload transport is calculated at each instant through an
  (asymmetric) wave cycle, based on the instantaneous wave-
  plus-current near-bed velocity applied to a bedload formula,
  and averaged over the wave cycle
- The suspended sediment concentration profile is deduced from
  a time-invariant eddy diffusivity, and is obtained by numerical
  integration of the concentration gradient through the vertical,
  taking account of high-concentration effects and density
  stratification.

- The velocity profile is calculated as two logarithmic profiles with different slopes inside and outside the wave boundary layer, matched at the edge of the boundary layer.
- The suspended load transport rate is calculated by a numerical integration through the vertical of the product of the time-mean concentration and velocity profiles.
- The total load transport rate is the sum of the bedload and suspended transport rates, and has a component in the wave direction as well as one in the current direction.

Although the method is not a numerical model in the conventional sense, it can only be implemented practically through use of a computer program. It is well supported by comparisons with data.

Danish STP method

This method is based on a numerical model of the wave-plus-current boundary layer with sediment diffusion. The numerical model uses a turbulent kinetic energy formulation with a grid of points through the vertical, and includes the effects of wave breaking.

The STP model consists of two parts: a hydrodynamic module and a sediment transport module.

The hydrodynamic module describes the turbulent boundary layer under combined waves and current, using the theory of Fredsøe (1984). The sediment transport module calculates the time-varying bed load and sediment concentration profile from the time-varying bed shear-stress and eddy-diffusivity profile obtained by the hydrodynamic module. The resulting bedload transport and suspended-load transport is found as the time average of the instanteous values. The sediment transport module is based on the work of Fredsøe et al. (1985) and Engelund and Fredsøe (1976).

For strong waves and for strong currents the bed is taken to be plane, but for weak waves and a weak current an option can be set to estimate the dimensions of wave ripples and their effect on the eddy diffusivity. The ripple predictions are based on empirical relations primarily obtained from the results of Nielsen (1979).

For ease of use, multiple runs of the model can be made for a discrete range of inputs for a particular problem to compile a multi-parameter computer-based look-up table. Sediment transport rates for any other inputs can be obtained with relative efficiency by a multi-parameter interpolation procedure.

The STP model is both versatile and sophisticated in terms of the physical process it includes. It is updated periodically to include new processes and features.

The STP model is just one example of a class of computer-based numerical models that have been developed in recent years to calculate the hydrodynamics, sediment dynamics, and sediment transport due to combined wave and current action at a point in an (assumed quasi-uniform) plane. Four of these models are described by Davies *et al.* (1997), namely those of A. G. Davies, J. S. Ribberink, A. Temperville and the STP model. They all use a grid of points in the vertical (finer near the bed), a grid of points in time finely dividing the wave cycle, some form of turbulent-energy closure scheme to calculate the time-varying eddy viscosity and diffusivity profiles, diffusion theory for sediment suspension, various degrees of flow-sediment interaction, and a finite-difference solution scheme. Two of the models use the computed bed shear-stresses to drive the bedload transport formula on an instant-by-instant basis. Davies *et al.* (1997) compared the performance of the four models with detailed laboratory data sets obtained in an oscillating water tunnel with regular waves alone, asymmetrical waves alone, and regular waves superimposed collinearly on a current, in plane bed sheet-flow conditions above a fine sand bed. Their performance compared with the measured concentration profiles is summarised in Section 8.6. When compared with the measured time-averaged sediment transport rates, all the models gave predicted rates within a factor of 2 of all the measured values.

Numerical models of this type probably represent the best way forward in developing more advanced, accurate, reliable sediment transport prediction methods. They can be used as versatile research tools to incorporate additional processes such as wave breaking, irregular asymmetric waves, mass transport velocities, bed slopes, density stratification, mixed sediments, etc. However, they are relatively inaccessible to practical users, are excessively demanding in computer time to be usable in morphodynamic modelling, and in many cases rely on the skill of their developer to obtain robust, reliable results. They also give less insight into the dependencies on input parameters than that given by analytical formulations. Interestingly, a comparison, by Deigaard (1998), between the STP model and the simple Bailard formula (Equation (134)) for a wide range of combinations of input parameters found surprisingly good agreement (within a factor of 3) between the two methods for a large proportion of the input combinations. Exceptions occurred for conditions in which the Bailard formula is not expected to be valid, namely near or below the threshold of motion, and for current dominated conditions. This suggests that the more sophisticated numerical models are not necessarily able to give significantly more accurate results than the simpler formula methods in many practical applications.

A promising way to make optimum use of the numerical models would be to use them to calibrate analytical methods such as the Bailard formula, or the Soulsby–Van Rijn formula (Equation (136)), or a more advanced and versatile version of such a formula. This would combine the physical detail obtainable with numerical models with the computational efficiency of analytical formulae.

All the formulae presented in this chapter are given in a uni-directional form. In the sea, two horizontal coordinates must be considered. Since the sediment transport is approximately aligned with the current (although wave effects can cause deviations of the direction of sediment transport by up to about $15°$ from the current direction), the sediment transport vector, $\boldsymbol{q}_t$, can be written as

$$\boldsymbol{q}_t = (q_{tx}, q_{ty})$$

$$= \left(|\boldsymbol{q}_t| \frac{U_x}{|\boldsymbol{U}|}, |\boldsymbol{q}_t| \frac{U_y}{|\boldsymbol{U}|} \right) \tag{137}$$

where

q_{tx}, q_{ty} = components of sediment transport rate in ortho-gonal x, y directions

$\boldsymbol{U} = (U_x, U_y)$ = current velocity vector

U_x, U_y = components of current in x, y directions

$|\boldsymbol{q}_t|, |\boldsymbol{U}|$ = magnitudes of sediment transport rate, current speed

The value of $|\boldsymbol{q}_t|$ is given by q_t from the formulae in the preceding Sections.

The formulae for q_t can be applied in a quasi-steady, instant-by-instant, form over a period of time (e.g. a tidal cycle, or a spring-neap cycle, or a year of storms), and the net transport obtained by integrating through time. Note that, because the sediment transport rate is proportional to a high power (3–5) of the current speed, the net sediment transport does not necessarily follow the residual current.

All these methods of calculating the sediment transport rate are based on assumptions that conditions are horizontally uniform, steady in time, and there is an inexhaustible supply of sediment. They are thus methods for calculating the equilibrium, or saturated, sediment transport. However, conditions in the sea and in estuaries often violate one or more of these conditions: the current speed, wave height or water depth may vary rapidly in space around headlands or sandbanks or in a sudden expansion of an estuary; the flow may be tidal rather than steady; or the flow may pass over areas of rock or inerodible gravel. Bedload transport responds quickly to such changes, and hence an approach based on quasi-equilibrium transport is justified. But suspended load transport responds much more slowly because it takes a finite time and distance for new sediment on the bed to be diffused up through the flow, and for sediment already in suspension to settle to the bed.

In such cases the sediment transport rate may be smaller (undersaturated) or larger (oversaturated) than the equilibrium value. The effects are most pronounced for fine grains, and short time or distance resolution. A rough criterion for the importance or otherwise of non-equilibrium effects is to compare the resolution length-scale L_R over which variations are of interest with the adjustment length L_A given by $L_A = 0.005\bar{U}^2 h/w_s^2$. If $L_R \leq L_A$, then non-equilibrium effects should be taken into account, whereas if $L_R \gg L_A$ they can be ignored. Taking an example of $h = 4\,\text{m}$, $\bar{U} = 1\,\text{m s}^{-1}$ and $w_s = 0.02\,\text{m s}^{-1}$, the adjustment length $L_A = 50\,\text{m}$. Thus, to resolve the details of sediment transport over sandwaves of wavelength $= 25\,\text{m}$ with a grid size of $L_R = 1\,\text{m}$, a non-equilibrium method must be used, but for a relatively coarse gridded coastal area model with a grid size of $200\,\text{m}$ an equilibrium model will be adequate.

Methods of calculating non-equilibrium transport have been devised, but full descriptions are outside the scope of this book. They range from schematised, computationally efficient, methods such as that of Miles (1981), to full numerical solutions of the convection–diffusion equations. An example of the latter applied to flow and sediment transport over sandwaves is given by Johns *et al.* (1990).

To improve the accuracy of sediment transport predictions for practical applications, it is often advisable to take field measurements in the study area and make a site-specific re-calibration of a sediment transport formula. Methods of measuring the transport include:

- For bedload, measure bedform migration rates, and convert to bedload transport rates (see Section 7.2). Alternatively, make measurements using radioactive or fluorescent tracers (Crickmore *et al.*, 1990).
- For suspended load, make simultaneous measurements of the velocity profile with a current meter and the suspended sediment concentration profile with a pumped sampling system (Crickmore and Teal, 1981). The suspended load transport rate can be calculated by making a numerical integration of the product of velocity and concentration through the water column.
- For channel infill problems, dredge a short trial trench and measure the infill rate by repeat echo sounding.

The corresponding input parameters, such as water depths, current speeds, wave conditions and sediment characteristics, must also be measured. The simplest method of re-calibrating a sediment transport formula is to adjust the leading coefficient in the formula; alternatively, and possibly more justifiably, the bed roughness can be adjusted. Other coefficients, such as the power to which current speed is raised or the wave stirring coefficient, can also be given site-specific values, but one should be suspicious if any of the re-calibrated coefficients differs greatly from the standard values. This could suggest either an error in the measured values, or that an inappropriate sediment transport formula has been chosen.

Procedure

1. To calculate the sediment transport rate q_t (volume/volume) for *combined waves and currents* outside the surf zone, obtain values

of the same quantities as for currents alone (Section 10.2), and the following additional quantities:

wave height	H
wave period	T
angle between waves and currents	ϕ

The procedure is illustrated using input values taken from Figure A.3 of Van Rijn (1993), which gives examples of the use of Van Rijn's **TRANSPOR** program. The same inputs are used below with the Soulsby–Van Rijn formula, and results are compared with the values from **TRANSPOR**.

Example 10.2. Total load transport by waves and currents

Obtain values of the following parameters:

water depth	h	5 m
median grain diameter	d_{50}	0·25 mm
90th percentile	d_{90}	0·50 mm
density of sediment grains	ρ_s	2650 kg m^{-3}
water temperature		15 °C
salinity		0 ppt
depth-averaged current speed	$\bar{U}$	0·6 m s^{-1}
significant wave height	H_s	1 m
peak wave period	T_p	6 s
angle between waves and current	ϕ	90°
slope of bed	β	0°
Calculate water density	ρ	999 kg m^{-3}
and kinematic viscosity	ν	$1·14 \times 10^{-6}$ m^2 s^{-1}
Calculate $s = \rho_s/\rho$	s	2·65
Calculate T_z from T_p, assuming a JONSWAP spectrum (Equation (48b))	T_z	4·67 s

Calculate RMS orbital velocity (see Example 4.3)	U_{rms}	0.262 m s^{-1}
For these relatively mild current and wave conditions the bed is expected to be rippled, and hence bed roughness is taken to be	z_0	0.006 m
Calculate drag coefficient due to current alone (Equation (37))	C_D	0.00488
Calculate threshold current speed using Van Rijn's method (Equation (71))	$\bar{U}_{cr}$	0.382 m s^{-1}
Calculate A_{sb} (Equation (136b))	A_{sb}	1.28×10^{-4} units
Calculate D_*	D_*	5.80
Calculate A_{ss} (Equation (136c))	A_{ss}	7.78×10^{-4} units
Calculate $A_s = A_{sb} + A_{ss}$	A_s	9.06×10^{-4} units
Calculate total sediment transport rate (Equation (136a))	q_t	60.6×10^{-6} m^2 s^{-1}
or in mass units, $\rho_s q_t =$	Q_t	0.161 kg m^{-1} s^{-1}

Comparing the values of A_{sb} and A_{ss}, it is seen that bedload and suspended load contribute 14% and 86% of the total transport rate, respectively. The same inputs but without waves ($H_s = 0$ m) yield $q_t = 14.1 \times 10^{-6}$ m^2 s^{-1}, so that the addition of waves increases the sediment transport by about a factor of 4 in this case. The value calculated by Van Rijn's TRANSPOR program for the same inputs is $q_t = 69.5 \times 10^{-6}$ m^2 s^{-1}.

2. Calculations similar to those in Example 10.2 have been made (using SandCalc) to produce a plot of q_t versus current speed for three wave conditions: $H_s = 0$; $H_s = 1$ m, $T_p = 6$ s; and $H_s = 3$ m, $T_p = 8$ s.
    In making these calculations to produce Figure 31, two constraints were relaxed: (a) all waves were assumed to be non-breaking; (b) $z_0 = 6$ mm was used in all cases, including ones

where ripples would probably be washed out.

Figure 31 shows how the presence of the waves amplifies the sediment transport rate by factors of 100 or more at low current speeds, but by progressively smaller factors for larger current speeds. At current speeds which would be below the threshold of motion on their own, the addition of waves allows substantial sediment transport to take place.

Also shown are the results presented by Van Rijn for the TRANSPOR method. They show broadly similar behaviour to the Soulsby–Van Rijn formula, although it must be borne in mind that one coefficient in the latter was calibrated against the TRANSPOR method. Differences between the two methods in the presence of waves are generally within about a factor of 2, and are not much greater than for the case $H_s = 0$ m corresponding to the difference between the current-alone Van Rijn formula and the full method.

The Soulsby–Van Rijn formula (Equation (136)) is recommended for its simplicity compared to the TRANSPOR program, although it should be used with caution in cases where the bed is not rippled. If in doubt, calculate $\theta_{max,s}$ (see Example 5.1) and check that it is less than 0·8 (see Equation (85a)).

10.5. LONGSHORE TRANSPORT

Knowledge

When waves break in the surf zone they release momentum, giving rise to a *radiation stress*. The cross-shore component of the radiation stress forces water onshore and causes a *set-up* of the water level, which rises in the onshore direction above the still water level. The water surface slope that this produces balances the cross-shore gradient of the shore-normal component of the radiation stress. For waves incident obliquely on the shoreline there is also a longshore component of the radiation stress, whose gradient gives rise to a *longshore current* within (and just outside) the surf zone which is balanced by friction with the bed. This in turn drives sediment along-shore as a *littoral drift*. The *longshore sediment transport rate* Q_{LS} measures the littoral drift across a normal to the shoreline. Variations in Q_{LS} along the shoreline cause recession or advance of the shoreline; in particular, if an obstruction such as a groyne, harbour entrance, or river training wall is built across the beach, the longshore transport causes sediment to build up on its updrift side and be eroded from its down-drift side.

Methods to calculate Q_{LS} are given below.

The CERC formula is the most widely used method to calculate the total sediment transport Q_{LS} integrated across the width of the surf zone. The formula, originally given in *Shore Protection Manual* (CERC, 1984) in US units, can be converted to a dimensionally consistent form (e.g. Fredsøe and Deigaard, 1992). A variety of versions have been used by practitioners, differing chiefly in their treatment of the wave group velocity c_g and the wave breaking criterion. Taking these as $c_g = (gh)^{1/2}$ and $H_{bk} = 0.8h$ in the surf zone, together with $H_s = \sqrt{2}H_{rms}$, leads to the simple formula:

$$Q_{LS} = \frac{0.023g^{1/2}H_{sb}^{5/2}\sin(2\alpha_b)}{(s-1)} \qquad \text{SC (138)}$$

where
$\quad Q_{LS}$ = sediment transport rate (m^3 s^{-1}) integrated across the surf zone, in volume of sediment (excluding pore space) per unit time
$\quad\quad g$ = acceleration due to gravity
$\quad H_{sb}$ = significant wave height at breaker line
$\quad\quad \alpha_b$ = angle between wave crests and shoreline at breaker line
$\quad\quad s$ = relative density of sediment.

Equation (138) is the simplest form of the CERC equation, obtained by applying shallow-water linear wave theory to the full expression. It appears to indicate that the transport rate Q_{LS} is independent of grain size and beach slope. Modifications to Equation (138) have been proposed by various researchers (e.g. Kamphuis, 1991; Ozasa and Brampton, 1980) to take account of:

- grain size of sediment (coefficient 0.023 becomes a decreasing function of grain size)
- beach slope (coefficient increases with beach slope)
- higher-order wave theories
- the along-shore variation in wave height (drives sediment from area of high waves to area of low waves).

The leading coefficient in the CERC formula was calibrated against longshore transport data taken from a number of sandy beaches with grain size smaller than about 0·6 mm, for which a large proportion of the sediment transport would be carried as suspended load. If the standard CERC formula is applied to larger grain sizes, such as shingle beaches, which are transported mainly as bedload, it is found that the predicted values are about 20 times larger than the observed values (Brampton and Motyka, 1984). It is therefore necessary either to re-calibrate the CERC formula, or use one of the modified versions, or use a formula specifically derived for coarse grain sizes.

Damgaard and Soulsby (1997) derived a physics-based formula for bedload longshore sediment transport. It is intended primarily for use on shingle beaches, although it is also applicable to the bedload component on sand beaches. The general principles are that the mean bed shear-stresses are calculated from the gradient of the radiation stress in the surf zone, and the oscillatory bed shear-stresses are calculated from the wave orbital velocities using a wave friction factor corresponding to the maximum of the values for rough turbulent flow using Equation (62b), and for sheet flow using an expression derived by Wilson (1989b). The resulting bed shear-stresses are applied directly in the Soulsby formula for bedload transport by combined waves and currents (Equations (129a–d)), without needing an intermediate step of calculating the longshore current distribution. A number of simplifying assumptions are made about the wave hydrodynamics in the surf zone, in order to make the mathematics tractable. The errors which these assumptions introduce are then corrected by re-calibrating the leading coefficient in the formula against a three-year field data set of measurements of longshore transport on a shingle beach in southern England. The resulting formula is:

$$Q_{LS} = \text{maximum of } Q_{LS1} \text{ and } Q_{LS2}$$

$$Q_{LS1} = \frac{0.19(g \tan \beta)^{1/2}(\sin 2\alpha_b)^{3/2} H_b^{5/2}(1 - \hat{\theta}_{cr})}{12(s - 1)}$$

$$\text{for } \hat{\theta}_{cr} < 1 \qquad \text{SC}\,(139a)$$

$$= 0 \qquad \text{for } \hat{\theta}_{cr} \geq 1 \qquad \text{SC}\,(139b)$$

$$Q_{LS2} = \frac{0.24f(\alpha_b)g^{3/8}d_{50}^{1/4}H_b^{19/8}}{12(s-1)T^{1/4}} \quad \text{for } \theta_{wr} \geq \theta_{wsf}$$

$$\text{SC}(139c)$$

$$Q_{LS2} = \frac{0.046f(\alpha_b)g^{2/5}H_b^{13/5}}{12(s-1)^{6/5}(\pi T)^{1/5}} \quad \text{for } \theta_{wr} < \theta_{wsf} \quad \text{SC}(139d)$$

$$\text{subject to } Q_{LS2} = 0 \text{ for } \theta_{max} \leq \theta_{cr} \quad \text{SC}(139e)$$

where

$$\hat{\theta}_{cr} = \frac{16.7\theta_{cr}(s-1)d_{50}}{H_b(\sin 2\alpha_b)(\tan \beta)}$$

$$f(\alpha_b) = (0.95 - 0.19\cos 2\alpha_b)(\sin 2\alpha_b)$$

$$\theta_{wr} = \frac{0.15H_b^{3/4}}{(s-1)g^{1/4}(Td_{50})^{1/2}}$$

$$\theta_{wsf} = \frac{0.0040H_b^{6/5}}{(s-1)^{7/5}g^{1/5}T^{2/5}d_{50}}$$

$$\theta_w = \text{maximum of } \theta_{wr} \text{ and } \theta_{wsf}$$

$$\theta_m = \frac{0.1H_b(\sin 2\alpha_b)(\tan \beta)}{(s-1)d_{50}}$$

$$\theta_{max} = [(\theta_m + \theta_w \sin \alpha_b)^2 + (\theta_w \cos \alpha_b)^2]^{1/2}$$

$\theta_{cr} = $ threshold Shields parameter (Equation (77))

$H_b = $ wave height at breaker line

$T = $ wave period

$d_{50} = $ median grain diameter

$\tan \beta = $ beach slope

and other quantities are defined after Equation (138). Compared with the CERC formula (Equation (138)), the Damgaard and Soulsby formula should have a wider applicability because it has as a stronger basis in physics, and it includes dependencies on grain size, beach slope and wave period which the standard CERC formula does not. However, the CERC formula includes suspended load transport, and has the virtue of simplicity.

Procedure

1. To calculate the longshore sediment transport for fine grain sizes (say less than 0·5 mm), which travel in suspension as well as by bedload, use the CERC formula (Equation (138)). For coarse grain sizes (> 0·5 mm), which travel mainly as bedload, use the Damgaard and Soulsby formula (Equation (139)).

2. In either case a representative wave height must be used. For the CERC formula, the leading coefficient given in Equation (138) is appropriate for the input of a significant wave height H_s. For the Damgaard and Soulsby formula, use $H_b = H_{rms} = H_s/\sqrt{2}$, and $T = T_p$, where H_{rms} is the root-mean-square wave height and T_p is the peak wave period (see Section 4.2).

3. The procedure to calculate the longshore transport rate on a shingle beach is illustrated using the Damgaard and Soulsby method.

Example 10.3. Longshore transport

Obtain values of the following parameters:		
median grain diameter	d_{50}	10 mm
significant wave height at breaker line	H_{sb}	1 m
peak wave period	T_p	6 s
angle between wave crests and shoreline at breaker line	α_b	10°
relative density of sediment	s	2·58
beach slope	$\tan \beta$	1/10
	$\Rightarrow \beta$	2·86°
Calculate RMS wave height at breaking $= H_{sb}/\sqrt{2}$	H_b	0·707 m
Take effective wave period $= T_P$	T	6 s
Calculate threshold Shields parameter using Soulsby formula (see Example 6.3)	θ_{cr}	0·0553

Calculate threshold factor

$$\frac{16\cdot7 \times 0\cdot0553(2\cdot58 - 1) \times 0\cdot01}{0\cdot707 \sin 20° \times (1/10)}$$

$$= \hat{\theta}_{cr} \qquad 0\cdot603$$

Calculate directionality factor

$$(0\cdot95 - 0\cdot19 \cos 20°)$$

$$\sin 20° = f(\alpha_b) \qquad 0\cdot264$$

Calculate rough-flow wave Shields parameter

$$\frac{0\cdot0040 \times 0\cdot707^{3/4}}{(2\cdot58 - 1)9\cdot81^{1/4}(6 \times 0\cdot01)^{1/2}}$$

$$= \theta_{wr} \qquad 0\cdot169$$

Calculate sheet-flow wave Shields parameter

$$\frac{0\cdot0040 \times 0\cdot707^{6/5}}{(2\cdot58 - 1)^{7/5} \times 9\cdot81^{1/5} 6^{2/5} \times 0\cdot01}$$

$$= \theta_{wsf} \qquad 0\cdot043$$

Take $\theta_w = \max(\theta_{wr}, \theta_{wsf})$ $\qquad \theta_w \qquad 0\cdot169$

Calculate mean Shields parameter

$$\frac{0\cdot1 \times 0\cdot707 \times \sin 20° \times (1/10)}{(2\cdot58 - 1) \times 0\cdot01}$$

$$= \theta_m \qquad 0\cdot153$$

Calculate maximum Shields parameter

$$[(0\cdot153 + 0\cdot169 \sin 10°)^2$$
$$+ (0\cdot153 \cos 10°)^2]^{1/2} \qquad \theta_{max} \qquad 0\cdot237$$

Check if $\hat{\theta}_{cr} < 1$, and
$\theta_{max} > \theta_{cr}$, otherwise the
flow is below threshold
and there is zero transport.
In the example, both are
satisfied.

Calculate Equation (139a) $\quad Q_{LS1}$

$$3\!\cdot\!31 \times 10^{-4}\ \mathrm{m^3\,s^{-1}}$$

Since $\theta_{wr} > \theta_{wsf}$, use
Equation (139c) $\quad Q_{LS2}$

$$6\!\cdot\!98 \times 10^{-4}\ \mathrm{m^3\,s^{-1}}$$

Take
$Q_{LS} = \max(Q_{LS1}, Q_{LS2}) \quad Q_{LS}$

$$0\!\cdot\!00698\ \mathrm{m^3\,s^{-1}}$$

If preferred, convert to
volume transport per year,
including pore-space

$$= \frac{0\!\cdot\!00698 \times 3600 \times 24 \times 365}{(1 - 0\!\cdot\!40)}$$

$$= 36\,700\ \mathrm{m^3\ year^{-1}}$$

4. For comparison, the CERC formula (Equation (138)) with the same inputs gives $Q_{LS} = 0\!\cdot\!0156\ \mathrm{m^3\,s^{-1}}$, or $819{,}410\ \mathrm{m^3/year^{-1}}$. This value is about 22 times larger than that given by the Damgaard and Soulsby method, and is probably an overestimate by this factor because the standard CERC formula is calibrated for suspended transport of fine sands rather than bedload transport of shingle.

5. Usually a series of several years of wave data must be used sequentially as inputs to the chosen formula to calculate long-term beach response. The results are often very sensitive to the orientation of the shoreline which is taken. An error of $1°$ in the assumed shoreline orientation can cause a 10% error in the net transport.

6. The above methods should not be used if there are tidal currents near the coast of speed greater than about 20% of the longshore current speed; nor if the beach is significantly non-planar; nor if the beach has zoned sediment sizes (e.g. shingle upper beach and sand lower beach). In these cases, a better solution is to use a Coastal Profile Numerical Model (see Section 11.1) to calculate

the hydrodynamics and the sediment transport point-by-point across the profile. Such models can also calculate the cross-shore distribution of longshore sediment transport, take account of the effect of cross-shore structures such as groynes, and take account of the effect of backing seawalls. An example is the numerical model COSMOS–2D (Southgate and Nairn, 1993; Nairn and Southgate, 1993).

Morphodynamics and scour

11

11. Morphodynamics and scour

11.1. MORPHODYNAMIC MODELLING

Knowledge

The study of the changes with time in the shape of the bed of a
river, estuary or sea is know as *morphodynamics*. When the
changes result from the presence of an object or structure, the
process is referred to as *scour*. The general subject area of
morphodynamics and scour will be treated only rather briefly in
this manual, as it is a large and complex subject which is treated
in detail in more specialised texts.

One of the commonest reasons for needing to consider sand
transport is to allow predictions to be made of areas of sediment
accretion or erosion (and their rates) at a coastal or estuarine
site. This is accomplished by means of the sediment budget
equation, which for one-dimensional applications can be written as

$$\frac{\partial \zeta}{\partial t} = -\frac{1}{1-\varepsilon}\left(\frac{\partial q_{\mathrm{B}}}{\partial x} + D - E\right) \tag{140}$$

where ζ = bed-level, relative to an arbitrary datum
x = horizontal coordinate
t = time
ε = porosity of bed
q_{b} = volumetric bedload transport rate in positive x
direction
D = deposition rate, expressed as volume of sediment
grains settling from suspension on to unit area of bed
per unit time
E = erosion rate, expressed as volume of sediment grains
eroded into suspension from unit area of bed per unit
time

The process expressed by Equation (140) is illustrated in Figure 32, for the case of bedload transport alone. Over large distances (e.g. 100 m) and times, the 'storage' of suspended sediment represented by the difference $(E - D)$ in Equation (140) can be represented by replacing the bedload transport rate q_b with the total load transport rate q_t, to give the following equation:

$$\frac{\partial \zeta}{\partial t} = -\frac{1}{1 - \varepsilon} \frac{\partial q_t}{\partial x} \tag{141}$$

where q_t = total (suspended and bedload) volumetric sediment transport rate in the positive x direction

The dependence of q_t on position (x) and time (t) can be obtained from one of the sediment transport formulae given in Chapter 10.

In two horizontal dimensions, Equation (141) can be written as

$$\frac{\partial \zeta}{\partial t} = -\frac{1}{1 - \varepsilon} \left(\frac{\partial q_{tx}}{\partial x} + \frac{\partial q_{ty}}{\partial y} \right) \tag{142}$$

where q_{tx}, q_{ty} = components of total volumetric sediment transport rate in positive x, y directions
y = horizontal coordinate orthogonal to x

For applications such as infill of dredged channels or trenches for pipelines and cables, and dispersal of spoil heaps or sediment storage heaps, it may be possible to make analytical morphodynamic calculations. However, for larger coastal and estuarine areas it is necessary to use a computational model. The essence of these models is that the shape of the bed is updated at regular intervals, and the wave, current and sediment transport patterns are re-computed with the new bathymetry. This process is repeated for the required duration.

There are three main classes of coastal morphodynamic numerical model:

- *Beach planshape models*, which compute the changes in position and shape of the coastline (e.g. the mean still-water level along a beach) over periods of years or decades. A long sequence of wave heights and directions is used as input. These are refracted in from deep water to the breaker line, and drive a longshore sediment transport formula such as the

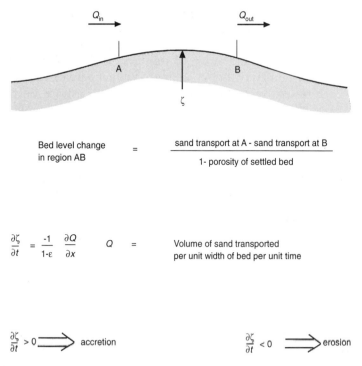

Bed level change in region AB $=$ $\dfrac{\text{sand transport at A - sand transport at B}}{\text{1- porosity of settled bed}}$

$$\frac{\partial \zeta}{\partial t} = \frac{-1}{1-\varepsilon} \frac{\partial Q}{\partial x}$$

Q $=$ Volume of sand transported per unit width of bed per unit time

$\dfrac{\partial \zeta}{\partial t} > 0 \implies$ accretion $\qquad$ $\dfrac{\partial \zeta}{\partial t} < 0 \implies$ erosion

Figure 32. Principle of morphodynamics

CERC formula at each of a grid of points along the coastline. The coastline recession or advance is calculated at each time step by computing the sediment budget (transport in minus transport out) for each cell lying between grid points. This results in an updated planshape of the coastline. The process is repeated for the duration of the available wave sequence. Enhancements include: using a more sophisticated transport formula, distributing the transport across the surf zone and using several lines along the coast instead of just one.

- *Coastal profile models*, which model the cross-shore beach profile along a normal to a straight or gently curving coastline. A grid of points is taken along the normal, waves are refracted in from deep water, break according to a breaking criterion, and continue to propagate through the surf zone. Some account is taken of variations in stresses and velocities through the vertical. Computations are made of the distribution across the profile of the wave height, set-up, bed shear-stresses, wave asymmetry, undertow velocity, and resulting cross-shore sediment transport rate (e.g. using the Bailard formula). The change in the shape of the beach profile at each time step is calculated from the sediment budget between grid points. This results in an updated shape of cross-shore bottom profile. The process is repeated for each successive wave condition. Enhancements include: a more sophisticated breaking scheme, a more sophisticated sediment transport formula, the inclusion of tidal rise and fall in water levels, the inclusion of tidal currents, and the inclusion of longshore as well as cross-shore sediment transport. Typically, a breaker bar is formed due to the convergence of onshore directed sediment transport seaward of the breaker point resulting from wave asymmetry, and offshore directed transport landward of the breaker point resulting from undertow. Coastal profile models are more computationally demanding than beach planshape models, so that shorter sequences of waves are usually used (several weeks or months) but they are also considerably more versatile.

- *Coastal area models*, which compute the pattern of erosion and accretion over a topographically complex sea area or estuary. They contain linked modules for wave propagation, current distribution (tidal, wave-driven, wind-driven), and the resulting sediment transport over a horizontally two-dimensional sea area. The morphodynamic evolution of the seabed is calculated by a two-dimensional sediment budget equation such as Equation (142). The sediment transport rates in two dimensions are calculated from methods such as the Soulsby–Van Rijn formula or the (schematised) Danish STP model. They are usually either depth-averaged (2DH), or contain some simple representation of variations through the vertical (Quasi 3D). They make greater demands on computer time than either of the other classes of model, and therefore are usually not run for longer than about 14 days. Special techniques must be used to extend the prediction period to years or decades. Fully three-dimensional models are also starting to appear, but at present they are excessively computer hungry.

Fuller details of these types of models and related morphodynamic issues can be found in review volumes edited by de Vriend (1993) and Stive *et al.* (1995). A comprehensive book on the subject is being compiled by de Vriend (in preparation).

Procedure

1. Obtain information on the bathymetry and sediment characteristics of the study area.

2. Obtain the pattern and time-history of waves by measurement, estimation (e.g. hindcasting from wind data) or from a computational model (e.g. refraction/shoaling/diffraction) of the study area.

3. Obtain the pattern and time-history of currents by measurement, estimation (e.g. from tidal diamonds on an Admiralty chart) or from a computational model of the study area. Tidal, wind-induced and wave-induced currents may need to be included.

4. Calculate the pattern and time-history of the sediment transport rate, using one of the formulae in Chapters 9 and 10.

5. Calculate the pattern of erosion and accretion rates by means of Equation (141) or (142).

6. This gives the *initial* pattern of erosion and accretion rates, which can be integrated over a period of time to give the new bathymetry. However, if the depth changes by more than about 10%, the wave and current patterns will change in response. For longer period calculations, return to step 1 with the new bathymetry and follow through steps 2–5. Repetition of this cycle gives a fully *morphodynamic* computation of the long-term response of the study area.

7. For changes in coastline position and shape use a beach planshape model, for cross-shore changes in beach profile use a coastal profile model, and for erosion/accretion patterns in complex topographies and further offshore use a coastal area model. All these numerical models need to be run by specialists experienced in their use.

11.2. SCOUR

Knowledge

When an object is placed on the seabed, the flow speeds up around it, so that more sediment is carried away from than is carried into the vicinity of the object. This causes a *scour pit* to form around the object.

Additional effects, such as secondary circulations and flow reversal, may also contribute to the scouring process.

Scour is a local form of morphodynamic response of the seabed, and many of the principles given in Section 11.1 apply. However, because the flow, and hence the sediment transport varies rapidly spatially, the non-equilibrium terms D and E in Equation (140) are important.

Examples of scour include: bridge piers; piles supporting jetties; toe-scour at sea-walls, harbour breakwaters, offshore breakwaters, etc; piles securing offshore oil platforms (jackets); gravity-base structures; oil and gas pipelines; spudcans of jack-up drilling rigs; burial of objects laid on the seabed; erodible beds adjacent to inerodible areas.

The scour may be caused by unidirectional currents, tidal currents, waves, or waves combined with currents.

Many aspects of scour are similar for all the objects and flows, but the detailed treatment of each is different.

Procedure

1. Prediction of the ultimate scour after a long period of wave or current flows around certain simple geometries of object can be made directly from existing knowledge.

   The scour pit around a vertical *cylindrical pile* in a steady current is roughly conical, having side-slopes of about 28°, and a maximum depth adjacent to the pile of about 1·4 times the pile diameter. The overall diameter of the scour pit is about six times the pile diameter. The above applies to piles which have a diameter more than 40 times the sediment grain size, and less than about 1/6 of the water depth, in a current whose speed exceeds the threshold of motion of the undisturbed bed.

   The scour pit beneath a horizontal circular pipe fixed so that its lower edge lies at the initial seabed level, in a steady current, has an ultimate maximum depth of about 0·6 pipe diameters.

The time development of the depth of scour occurs rapidly initially, tending asymptotically with time to its ultimate value according to an approximately exponentially decaying time dependence.

2. For less simple geometries of flows, a physical model test is the best solution. The following scaling laws apply:

   - ensure grain size $< 1/40$ of principal dimension
   - scale water depth with model scale, unless prototype depth is more than six times principal dimension of object, in which case model depth need only be more than six times dimension of object
   - current speed U/U_{cr} should be kept constant between model and prototype
   - scour pit dimensions scale directly with model scale
   - time-scaling laws are available (see Whitehouse, 1997).

3. Numerical modelling of scour is just about becoming practicable, although a number of problems need to be solved, and computing times are large. For the next few years, physical modelling is probably a better solution for most practical problems than numerical modelling. However, in the future numerical modelling may provide the most efficient solution for scour problems.

   A comprehensive guide to scour around marine structures is given by Whitehouse (1997).

Handling the
wave-current climate

12

12. Handling the wave-current climate

12.1. GENERAL

In the preceding sections it has been assumed that a single current speed and/or a single wave condition (or spectrum of waves) has been specified. However, for many real applications one is faced with wave and current conditions which vary enormously during a year, ranging from calm conditions coupled with neap tides to extreme storms coupled with peak spring tidal currents. The question then is how to make a prediction of average sediment transport (or morphodynamics or scour) which includes the effects of all these conditions correctly. The decision of what wave and current inputs to use is particularly important for calculation of long-term sediment transport or morphodynamics.

The question of handling the wave-current climate arises in connection with a wide variety of classes of application:

- stability of scour protection material around offshore structures
- development of scour around (unprotected) structures
- burial of objects lying on the seabed
- morphodynamic development of beaches and offshore areas with or without structures
- sediment mobility and transport pathways in offshore areas
- annual sediment ingress to water intakes
- annual infill rates of trenches, navigation channels and harbours.

Most of the basic principles described below are common to some or all of the above applications.

There are three basic approaches that may be used:

- design wave and tide approach
- probabilistic approach
- sequential approach.

All have their strengths and weaknesses, and are described below.

12.2. DESIGN WAVE AND TIDE APPROACH

In this approach, a single tidal height and current, and a single wave condition, are specified. In some studies these are specified by the client as the design conditions which a structure or scheme has to withstand. These may be based on parameters having structural significance, such as the 50-year storm and Highest Astronomical Tide conditions which might cause a structure to fail, but they are not necessarily the conditions which are most appropriate for sediment transport purposes, since the extreme condition may occur so rarely or be so short- lived that it has only a small effect on the long-term sediment behaviour. A structural fatigue condition which takes account of the cumulative behaviour over long periods is more nearly matched to sediment transport behaviour, although this is still not ideal.

A more appropriate choice of current and wave can be made by taking the values which give the greatest contributions to the long-term mean sediment transport, based on the probabilistic approach described in Section 12.3. As a rough guide, in tidal waters these are:

- peak current of mean spring tides, if waves are negligible
- peak current of mean tide (e.g. M_2), coupled with the wave having a 10% exceedance, if waves are important.

For longshore sediment transport calculations a rough guide is to take:

- the wave having a 20% exceedance.

In areas dominated by wind-driven currents, the design current and design wave should be linked via the wind-speed.

Better choices can be made if the probability distributions of current speed, $p_c(U_i)$, and wave height, $p_H(H_j)$, are known. Here, U_i is the current speed at the centre of the ith interval, and H_j is the wave height at the centre of the jth interval, of the probability histograms. Then the design current, U_d, and wave height, H_d, are given by:

$$U_{\mathrm{d}} = \left[\sum_i U_i^m p_{\mathrm{c}}(U_i) \right]^{1/m} \qquad (143a)$$

$$H_{\mathrm{d}} = \left[\sum_j H_j^n p_{\mathrm{H}}(H_j) \right]^{1/n} \qquad (143b)$$

The values of the powers m and n depend on the problem in hand: for bed shear-stress, $m = n = 2$; for bedload transport, $m = n = 3$; for total-load transport, m and n are in the range 3–5; for longshore transport, $n = 2\cdot5$.

More advanced techniques for choosing a 'morphodynamic tide' (Latteux, 1996) or a 'single representative wave' (Chesher and Miles, 1992) for detailed morphodynamic modelling have been developed to reduce computing time to manageable proportions.

It is important to be aware that the answer obtained in a sediment transport calculation will depend crucially on the choice of input current and wave conditions, and this choice is often a more important consideration than the choice of method used, or the inadequacies of sediment transport predictors.

Thus the design wave and tide approach should only be used if it is impossible or impractical to use the probabilistic or sequential approaches.

12.3. PROBABILISTIC APPROACH

It is often said that sediment transport is dominated by extreme events. This is only partly true. An extreme storm coupled with peak spring tidal currents will make a large impact but occurs very rarely, whereas moderate waves coupled with average tides occur very frequently. Both types of event may make similar contributions to the long-term sediment transport. The probabilistic approach includes all possible combinations of events summed with a weighting which depends on their frequency of occurrence.

Consider the case in which the currents are predominantly tidal. A probability distribution of the (say, hourly) current speeds can be obtained from either current meter measurements or a numerical model of the tidal currents. Denote this by $p_c(U_i)$, where p_c is the probability of occurrence of currents in the ith interval of speed centred on a speed U_i. Where neither measurements nor models are available, the probability distribution can be synthesised by combining charted hourly neap and spring tidal streams with a year's worth of tidal ranges from a tide table.

The wave climate (measured preferably over one year or more) can be represented by an $H_s - T_z$ scatter plot (Figure 12). Alternatively, it can be synthesised by hindcasting the waves from a long sequence of wind data. The RMS bottom orbital velocity is calculated (e.g. by linear wave theory, Section 4.4) for each combination of H_s and T_z. The mean water depth can be used in this calculation, provided that the tidal range is small compared with the mean water depth. The number of occurrences in the $H_s - T_z$ scatter plot falling into each interval of U_{rms} (e.g. $0 \cdot 1 \, \text{m s}^{-1}$ intervals) are then totalled and normalised by dividing by the total number of occurrences to yield the probability distribution of the RMS orbital velocity $p_w(U_{rms,j})$.

Since the tidal currents are generated by astronomical processes, and the waves are generated by meteorological processes, they can be regarded as being statistically independent of each other. With this important assumption, the joint probability of occurrence of a current U_i and a wave orbital velocity $U_{rms,j}$ is

$$p_{cw}(U_i, U_{rms,j}) = p_c(U_i)p_w(U_{rms,j}) \qquad (144)$$

Taking the sediment transport rate as an example of the approach, denote the sediment transport rate which results from a combination of a current U with a wave orbital velocity U_{rms} by $q(U, U_{rms})$. This can be calculated for a given grain size, etc., by the Soulsby–Van Rijn formula (Equation (136)), for example. The long-term average of the (gross) sediment transport rate $\langle q \rangle_{LT}$ is then given by

$$\langle q \rangle_{LT} = \sum_i \sum_j p_{cw}(U_i, U_{rms,j})q(U_i, U_{rms,j}) \qquad (145)$$

Thus a combination of Equations (144) and (145) can be calculated to give the long-term (e.g. annual) mean gross sediment transport rate at a point. This could be applied at every grid point of a coastal area model (Section 11.1) to give the distribution of the mean annual sediment transport over an area.

An extension of the above method is to subdivide the currents by direction (e.g. every 30°), and repeat the calculations given above for each direction class. The calculation given by Equation (145) then gives a long-term mean sediment transport rate and direction in every class of direction, which can be resolved into x and y components $\langle q_k \rangle_{LT,x}$ and $\langle q_k \rangle_{LT,y}$ for the kth class of direction. The long-term mean *net* sediment transport vector is then given by

$$\langle q \rangle_{LT} = \left(\sum_k \langle q_k \rangle_{LT,x}, \ \sum_k \langle q_k \rangle_{LT,y} \right) \tag{146}$$

This can be used to calculate long-term mean sediment pathways and magnitudes.

The gross sediment transport rate (Equation (145)) is the mean modulus of q ignoring direction, whereas the net sediment transport rate (Equation (146)) is the vector mean of q including direction.

Similar methods can be used for other categories of problem.

For scour and seabed mobility problems, denote the maximum bed shear-stress due to a current U_i and a wave orbital velocity $U_{w,j}$ by $\tau_{max}(U_i, U_{w,j})$. This should be the skin-friction component, and can be calculated by one of the methods described in Section 5.3. The cumulative probability of obtaining a shear-stress exceeding a particular value, say τ_{cr}, is obtained by summing the joint probabilities p_{cw} given by Equation (144) of all the combinations of U_i and $U_{w,j}$ which give a τ_{max} that exceeds τ_{cr}. If τ_{cr} is the threshold shear-stress for stability of a proposed scour protection, then the above calculation gives the annual probability of the protection being disturbed. In a seabed mobility problem, if τ_{cr} is the threshold of motion of the bed material, the calculation gives the proportion of the year for which the bed is mobile.

For sediment ingress problems, denote the concentration of suspended sediment at the height of the intake due to a current U_i and a wave orbital velocity $U_{w,j}$ by $C(U_i, U_{w,j})$. This can be calculated by the methods described in Section 8.6. Then the mean annual concentration of suspended sediment at the intake is given by

$$\langle C \rangle_{\mathrm{LT}} = \sum_i \sum_j p_{\mathrm{cw}}(U_i, U_{\mathrm{w},j}) C(U_i, U_{\mathrm{w},j}) \qquad (147)$$

The annual sediment load is given by multiplying $\langle C \rangle_{\mathrm{LT}}$ by the volume of water passing through the intake in one year.

For the methods given above it is necessary to calculate an equivalent monochromatic bottom orbital velocity U_{w} rather than the RMS value U_{rms}, because the corresponding equations for bed shear-stresses and suspended sediment concentrations are given in terms of monochromatic waves. In these cases, make the conversion

$$U_{\mathrm{w}} = \sqrt{2} U_{\mathrm{rms}} \qquad (148)$$

Additional refinements can be made as logical extensions of the above methods. These include making calculations for specific water levels correlated with current speeds in cases where the tidal range is not small compared with water depth. Another example is calculation of sediment ingress volumes subdivided by grain-size class.

Perhaps the most critical assumption made in the above method is that of the independence of the currents and waves used in Equation (144). While this is a fairly good assumption for areas with dominant currents such as much of the north-west European shelf, it will not be valid in areas where wind-induced or surge currents dominate, such as the northern North Sea and much of the North American shelf. It will also not be valid in the surf zone where wave-induced longshore currents dominate.

Even in tidally dominated areas the most extreme currents will have a wind-induced component. If U and U_{rms} are partially correlated, their joint probabilities become much larger for extreme events, so that these make a bigger contribution to the long-term mean. If the current and waves are completely correlated then the problem is actually simplified, because a single probability distribution (of wave height or wind speed) is sufficient. Cases of partial correlation (especially with a directional dependence) are much more complicated. In these cases a sequential approach, or a statistical approach based on a limited number of realisations, may be necessary.

Further details of the probabilistic approach are given by Soulsby (1987b) for gross sediment transport rates.

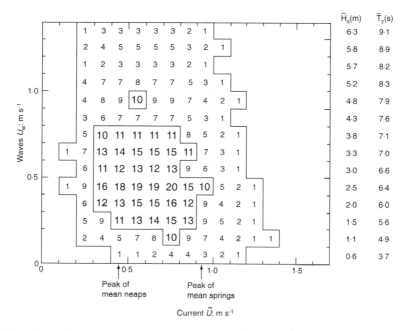

Figure 33. Contributions in ppt made by wave and current combinations to the long-term mean sediment transport

An example of the application to a site in the southern North Sea where long data sets of currents and waves were available was given by Soulsby (1987b). An illustration of this example is shown in Figure 33, in which the contributions (in parts per thousand) made by each combination of wave and current are shown. It is seen that large contributions (>10 ppt) are made by currents lying between the peaks of mean neap and mean spring tides, combined with waves in the range $1 \cdot 5 < H_s < 3 \cdot 8$ m. The latter correspond to exceedances in the range 33%–2%. The very largest contributions (19–20 ppt) are made by currents mid-way between the peaks of mean neap and spring tides, combined with waves of $H_s = 2 \cdot 5$ m, corresponding to an exceedance of 9%.

12.4. SEQUENTIAL APPROACH

The probabilistic approach uses the correct distributions of current and waves as inputs to sediment transport problems, but takes no account of the sequence in which they occur. In some types of application, especially morphodynamic problems, the sequence (or *chronology*) may be important. In this approach, the current (or surface elevations) and waves must be input at every timestep through a long-term simulation.

One example is beach planshape computations (see Section 11.1), where the formation of a feature in the shoreline may influence the subsequent pattern of wave transformation. As beach planshape models are relatively simple, it is computationally feasible to run them for a 30-year sequence of hourly wave inputs.

Coastal profile models can also be run for relatively long sequences. Southgate (1995*b*) took a four-month sequence of hourly wave data, split it into five segments, and re-ordered it into 120 feasible sequences. These were used as inputs to the coastal profile model COSMOS–2D with an initially plane beach. The resulting 120 final profiles showed large variations in shape, which did not depend strongly on the final segment of data encountered, illustrating that the chronology is important.

Numerical models of mud morphodynamics are particularly sensitive to chronology, in this case the sequence of tides, because of the history effect produced by consolidation of the bed. This produces different mud behaviour for tides increasing from neaps to springs to those which are decreasing from springs to neaps.

A difficulty with the sequential approach is that the chronology of wave inputs over a future period of years is not known *a priori*. Either a previous measured sequence must be used, or a synthetic time series generated with the correct statistical wave climate and a plausible sequence of events. This yields one realisation of the final morphology. This can then be analysed in terms of trends, variances and extremes. Ideally, a large number ($\approx$30) of different realisations should be made, and the results analysed statistically. Results must then be presented in statistical, or risk, terms.

The above procedure raises questions about: how to generate the input sequences in a physically realistic manner; whether the models developed for relatively short-term applications work correctly over long terms; and how to analyse and present the results.

The amount of computer time might also be prohibitive, especially for consultancy studies with tight budgets. In some cases the computer time required is completely impractical: for example, to make 30 runs of 10-year sequential predictions using a fine-gridded coastal area morphodynamic model would (at the time of writing) take about 100 years of continuous computing on a fast workstation! These and allied problems are presently at the frontiers of research. They are discussed in greater depth in the state-of-the-art book on morphodynamics by de Vriend (in preparation).

Case Studies

13

13. Case studies

13.1. GENERAL

Some case studies are presented in this chapter, to illustrate how the techniques described in the preceding chapters are used and combined to solve practical problems. The three examples are representative of some of the most commonly occurring types of application. The examples have been simplified and idealised to illustrate the main points involved, without introducing the site-specific complications which often occur in real studies, and which may require considerable ingenuity to overcome. The use of actual studies would have made the description of each case several times longer, without giving much additional general insight.

The calculations in the studies are referenced to equations and figures in the relevant chapters. However, the simplest method of performing many of the calculations is by means of the SandCalc software package, and the necessary steps to do this are described. In all the case studies, for simplicity the water is taken to have a temperature of 10 °C and a salinity of 35 ppt, and the sediment a density of 2650 kg m^{-3} (the default values in SandCalc). Other values can easily be used through the 'Edit-Water' and 'Edit-Bed Material' menus of SandCalc, using 'Derive'.

Each case study includes a sub-set of the steps outlined in the 'General Procedure' (Section 1.5), supplemented with case-specific procedures.

13.2. STABILITY OF SCOUR PROTECTION

An oil production platform is to be installed within an offshore oil field in a depth of 20 m at a site with a fine sand seabed. The eight 2·0 m diameter, circular cross-section, legs of the platform penetrate the bed. To avoid scour occurring, a protection blanket of rock will be placed around the legs. The size of rock necessary to provide protection against significant damage with a 50-year return period (for example) must be calculated.

The 50-year storm wave conditions at the site are specified as having $H_s = 7$ m. The effect of tidal currents is neglected. Assume that significant damage will occur if the 1/3 of highest waves in a given wave record are capable of moving the rock. Thus the critical condition corresponds (from the definition of H_s) to monochromatic waves with a design height of $H = 7$ m. The corresponding wave period is taken to be the peak period, T_p, of the spectrum. Using either Equations (49) and (48b), or SandCalc-Edit-Waves-Derive with $h = 20$ m (set in Edit-Water), $H_s = 7$ m, gives estimated values of $T_z = 9·3$ s and $T_p = 11·9$ s. The bottom orbital velocity amplitude, using linear wave theory, can be calculated from the monochromatic curve in Figure 14, or by using SandCalc Hydrodynamics–Orbital Velocity–Mono-chromatic with inputs of water depth = 20 m, wave height = 7 m, period = 11·9 s to give wave orbital velocity = 1·98 m s^{-1}. Experiments by Sumer et al. (1992) showed that the bed shear-stress due to wave motions around a vertical circular cylinder of diameter D has an amplification factor, due to the speed-up of the flow around the cylinder, which is a function of the Keulegan–Carpenter number, KC. In the present case, for $KC = U_w T/D = 1·98 \times 11·9/2·0 = 11·9$, they obtained a bed shear-stress amplification factor of 2·2. This can be interpreted (see Equation (57)) as being caused by an amplification factor of the orbital velocity of $(2·2)^{1/2} = 1·48$. Thus the effective orbital velocity adjacent to the platform leg is $1·48 \times 1·98 = 2·94$ m s^{-1}.

Anticipating that the size of rock required will be larger than 10 mm, the critical diameter d_{cr} can be obtained from Equation (79), with $U_w = 2·94$ m s^{-1}, $T = 11·9$ s, g = 9·81 m s^{-1}, and $s = 2·58$:

$$d_{cr} = \frac{97·9 \times 2·94^{3·08}}{11·9^{1·08}[9·81(2·58 - 1)]^{2·08}}$$

$$= 0·62 \text{ m}$$

Thus rock of diameter 0·62 m (mass approximately $= \pi \times 0·62^3 \times 2650/6 = 0·34$ tonnes) will not suffer significant damage in less than the 50-year return period storm conditions. In fact, since the velocity multiplier around a circular cylinder decreases rapidly with distance from the cylinder, a rather smaller rock would probably be sufficient.

Additional factors that might be required for specific studies include:

- calculating d_{cr} for a more stringent degree of acceptable damage, e.g. damage by the 1% of highest waves. The 1% exceedance wave height (approximately $1.5H_s$) can be calculated by assuming a Rayleigh distribution of wave height (see CIRIA/CUR, 1991)
- taking account of other structures, in the vicinity of the leg, such as a pipeline, or the temporary presence of a jack-up rig or a semi-submersible crane vessel. The velocity multiplier calculated for the other structure is multiplied by the multiplier (1.48) for the leg when calculating the effective orbital velocity
- in areas with strong currents (e.g. the southern North Sea), the combined effects of waves and currents may need to be taken into account by calculating the maximum shear-stress by the method described in Section 5.3. The calculation of return period of the stress is then an exercise in joint probability, as described in Section 12.3
- a filter layer of smaller material is usually laid beneath the main layer of rock, to prevent sand being winnowed out through the interstices of the rocks.

The manual *Scour at marine structures* (Whitehouse, 1997) gives a detailed description of scour protection methods.

13.3. INGRESS OF SEDIMENT TO A WATER INTAKE

A cooling water intake for a power station is to be sited offshore above a sandy seabed. The annual sediment load drawn in with the water has to be calculated, so that precautions such as provision of a sediment settling basin can be taken if necessary to prevent sediment entering the works. The mean water depth at the site of the intake is 7 m, the height of the centre of the intake is 3 m, the tidal range is 2 m, and the bed is a well-sorted fine sand with $d_{10} = 0.07$ mm, $d_{50} = 0.12$ mm and $d_{90} = 0.18$ mm. The currents in the area are predominantly tidal with a peak current speed on neap tides of 0.4 m s^{-1} and on spring tides of 0.6 m s^{-1}. The waves have a 50% exceedance height of $H_s = 0.5$ m, a 10% exceedance of $H_s = 1.0$ m and a 1% exceedance of 3 m. The mean direction of the wave approach is at 45° to the current direction.

The effects of both the current and the waves are likely to be important in suspending the sediment, and over a year a wide range of combinations will occur. For illustration purposes, a single value of current and wave is used, namely the peak current of mean tides, $U = \frac{1}{2}(0.4 + 0.6) = 0.5$ m s^{-1}, and the 10% exceedance wave height, $H_s = 1.0$ m. The wave period is not given, so an estimate is obtained using Equation (49), giving $T_z = 3.51$ s.

The formula to be used for the concentration profile under combined waves and currents is Equation (115a–e), which requires a monochromatic wave input. The equivalent monochromatic wave for these purposes is given by $H = H_s/\sqrt{2} = 0.707$ m s^{-1} and $T = T_p = 1.28 T_z = 4.50$ s. The bottom orbital velocity amplitude obtained from Figure 14 (see Example 4.2) is $U_w = 0.225$ m s^{-1}. The mean and maximum bed shear-stresses are obtained by using the method given in Section 5.3 with the 'DATA 13' coefficients from Table 9, and using a grain-related roughness of $z_0 = d_{50}/12 = 0.01$ mm, to give $\tau_m = 0.285$ N m^{-2} and $\tau_{max} = 0.481$ Nm^{-2}.

Next, the median grain size of the suspended sediment is calculated using the Van Rijn method, Equation (97). This method, which is really intended just for currents, requires the skin-friction shear-stress for currents which can be calculated using Equation (34) to give $\tau_{0s} = 0.228$ N m^{-2}, and the threshold bed shear-stress which can be calculated from Equation (76) to give $\tau_{cr} = 0.151$ N m^{-2}. It also requires d_{16} and d_{84}, which can be calculated by a log-normal interpolation (Section 2.2) between the known values d_{10}, d_{50}, d_{90} to give $d_{16} = 0.0793$ mm and $d_{84} = 0.164$ mm. These in Equation (97) give the median suspended grain size, $d_{50,s} = 0.106$ mm. The settling velocity of this grain size is given by Equation (102), $w_s = 0.00633$ m s^{-1}.

All the above quantities can be used in Equation (115), with $z = 3$ m, to give the concentration at the centre of the intake as $C(z) = 6.38 \times 10^{-6}$ by volume, or 16.9 mg l^{-1} by mass.

These calculations can alternatively be made using SandCalc as follows:

h	Edit-Water, set $h = 7$ m
d_{16}, d_{84}	Edit-Bed Material-Derive, with d_{10}, d_{50}, d_{90} entered
H, T	Edit-Waves-Derive, with $H_s = 1{\cdot}0$ m
U_w	Hydrodynamics–Waves–Orbital velocity–Monochromatic
τ_m, τ_{max}	Hydrodynamics–Waves + Currents–Total Shear-stress–Soulsby 13
τ_{0s}	Hydrodynamics–Currents–Skin-friction–Soulsby
τ_{cr}	Sediments–Threshold–Bed shear-stress–Soulsby
$d_{50,s}$	Sediments–Suspension–Suspended Grain size–Van Rijn
w_s	Sediments–Suspension–Settling velocity–Soulsby
$C(z)$	Sediments–Suspension–Concentration–Waves + Currents

Note that:

(a) the suspended grain size would be a little coarser if the wave effects are included in its calculation; for example, if $d_{50,s} = d_{50,b} = 0{\cdot}12$ mm, this would reduce the concentration to only $1{\cdot}9$ mg l^{-1}

(b) the equation used to calculate the concentration profile is designed for flat-bed, sheet-flow conditions, whereas the bed is more likely to be rippled for the wave and current used. A rippled bed would increase the concentration at $z = 3$ m.

To calculate the annual sediment ingress, it is necessary to take the following additional steps:

• make the above calculations for every combination of tidal current and wave condition occurring in a typical year, and sum the concentrations weighted by their frequency of occurrence, using the probabilistic method described in Section 12.3, to obtain an annual mean concentration

• multiply the annual mean concentration of sediment by the volume of water passing through the intake in a year, to give the annual sediment ingress.

Additional factors which may need to be taken into account include:

• the calculation is very sensitive to water temperature, because the reduced viscosity gives a larger settling velocity; an increase of temperature from $10\,°C$ to $30\,°C$ reduces the concentration at $z = 3\,m$ from $16\cdot9\,mg\,l^{-1}$ to $0\cdot20\,mg\,l^{-1}$.

- if the bed sediment is widely graded, possibly including a fine silty fraction and/or a coarse non-suspendible fraction, then a method which deals with a range of grain sizes must be used
- if only grain sizes coarser than a given size are of concern to the operation of the plant (e.g. if very fine sediment passes straight through), then a calculation subdivided by grain size must be made
- because of these sensitivities, some on-site measurements of suspended sediment concentration are valuable for calibration and validation purposes.

In addition to power stations, prevention of sediment ingress is important for water intakes to refineries and desalination plants.

13.4. INFILL OF A DREDGED CHANNEL

An approach channel to a harbour is to be dredged through a sandy shoal, and the annual maintenance dredging requirement has to be calculated. The channel is aligned at right angles to the coastline, such that tidal currents pass over it at right angles, and the dominant wave direction is aligned along the channel. The mean water depth at the shoal is 5 m and the channel is dredged to a mean navigation depth of 10 m, with gently sloping sides and width of 100 m at the deepest part. The tidal range is 2 m, tidal currents are equal on flood and ebb with peak speeds of $0\cdot6\,m\,s^{-1}$ on neaps and $1\cdot0\,m\,s^{-1}$ on springs. The waves have a 50% exceedance of $H_s = 1\cdot0\,m$ a 10% exceedance of $H_s = 2\cdot5\,m$, and 1% exceedance of $H_s = 2\cdot5\,m$. The sand on the shoal is well-sorted, with $d_{10} = 0\cdot1\,mm$, $d_{50} = 0\cdot2\,mm$, $d_{90} = 0\cdot3\,mm$.

The infill mechanism (in simplified terms) is caused partly by the currents, and partly by the waves. The current carries sediment across the shoal and into the channel, where the current is slower (because of continuity of mass) in the greater depth. The slower current carries less sediment out of the far side of the channel, so that there is net deposition of sediment in the channel. The process of more sediment being carried into than out of the channel occurs on both the ebb and flood currents, and their symmetry means that there is no net migration of the channel. Waves are assumed to have the same height and period inside the channel as outside on the shoal (i.e. refraction is ignored for the simplified description). The wave orbital velocities enhance the sediment transport rates, but because the orbital velocities are larger on the shoal than in the deeper water of the channel, the enhancement of the sediment transport is greater on the shoal. Thus the waves not only speed up, but also reinforce, the infill mechanism due to the currents.

As an illustration, the infill rate for a single current speed and wave height will be calculated. The values chosen are the peak current speed of the mean tide, $\bar{U} = \frac{1}{2}(0 \cdot 6 + 1 \cdot 0) = 0 \cdot 80$ m s^{-1}, and the 10% exceedance wave height, $H_s = 1 \cdot 0$ m. The corresponding zero-crossing period is not given, so it is calculated from Equation (49) as $T_z = 3 \cdot 51$ s. The water depth is taken to the mean level, $h = 5$ m outside, and 10 m inside the channel, as the tidal range is relatively small.

First, calculate the sediment transport rate, q_{in}, due to combined waves and currents on the shoal, using the Soulsby–Van Rijn formula (Equation (136)). The bed is assumed to be rippled sand with $z_0 = 6$ mm, and the current-only drag coefficient C_D is calculated from Equation (37) with inputs $h = 5$ m, $z_0 = 0 \cdot 006$ m, giving $C_D = 0 \cdot 00488$.

The RMS wave orbital velocity is obtained from Figure 14 using the JONSWAP curve (see Example 4.3), with inputs $H_s = 1 \cdot 0$ m, $T_z = 3 \cdot 51$ s, $h = 5$ m, giving $U_{rms} = 0 \cdot 208$ m s^{-1}.

The threshold current speed is calculated from Equation (71) with $h = 5$ m, $d_{50} = 0 \cdot 2$ mm, $d_{90} = 0 \cdot 3$ mm, giving $\bar{U}_{cr} = 0 \cdot 391$ m s^{-1}.

The sediment transport rate is calculated from Equation (136) with $h = 5$ m, $\bar{U} = 0 \cdot 8$ m s^{-1}, $C_D = 0 \cdot 00488$, $U_{rms} = 0 \cdot 208$ m s^{-1}, $\bar{U}_{cr} = 0 \cdot 391$ m s^{-1}, $\beta = 0°$ (horizontal bed), giving $q_t = 1 \cdot 838 \times 10^{-4}$ m^2 s^{-1}.

Now calculate the sediment transport rate in the trench, where $h = 10$ m. The current speed over the trench can be calculated from the equation of continuity: $\bar{U}_1 h_1 = \bar{U}_2 h_2$, giving $\bar{U} = 0 \cdot 8 \times 5/10 = 0 \cdot 4$ m s^{-1}. A series of calculations like those over the shoal, but using $h = 10$ m, $\bar{U} = 0 \cdot 4$ m s^{-1} give $C_D = 0 \cdot 00388$, $U_{rms} = 0 \cdot 083$ m s^{-1}, $\bar{U}_{cr} = 0 \cdot 416$ m s^{-1} and $q_t = 5 \cdot 28 \times 10^{-8}$ m^2 s^{-1}.

The above calculations, first over the shoal and then in the trench, can alternatively be performed using SandCalc, as follows:

h	Edit-Water, set $h = 5$ m
T_z	Edit-Waves-Derive, with $H_s = 1{\cdot}0$ m
C_D	Hydrodynamics–Currents–Total shear-stress–Logarithmic
U_{rms}	Hydrodynamics–Waves–Orbital velocity–Spectrum
U_{cr}	Sediments–Threshold–Current–Van Rijn
q_t	Sediments–Total load–Waves & Currents–Soulsby/Van Rijn
	Return to Edit-Water, set $h = 10$ m, and repeat the calculations

The net sediment transport rate is thus $1{\cdot}838 \times 10^{-4} \times 5{\cdot}28 \times 10^{-8} = 1{\cdot}837 \times 10^{-4}$ (the transport out of the trench is very small in this case). The sediment is deposited across the 100 m width of the trench, with a mean infill rate per hour, allowing for the pore space of the settled deposit with $\varepsilon = 0{\cdot}40$, of

$$\frac{1{\cdot}837 \times 10^{-4} \times 3600}{100(1 - 0{\cdot}40)} = 11 \ \text{mm h}^{-1}$$

This takes place only at the peak of the tidal current, and the infill rate at other times in the tide will be much less, so the total averaged over a tide will be smaller.

To calculate the annual maintenance dredging requirement it is necessary to take the following additional steps:

- make the above calculations for every combination of tidal current and wave condition occurring in a typical year, and sum the sediment transport rates weighted by their frequency of occurrence, using the probabilistic method described in Section 12.3, to obtain an annual mean infill rate
- perform the calculations at a number of representative points along the channel, and add the infills from each section.

Additional factors which may be important in channel infill calculations include:

- wave refraction over the channel; under certain conditions this can cause 'total internal reflection' of obliquely incident waves
- deflection of obliquely incident currents, due to the slowing effect as the current crosses the channel
- steep side-slopes may cause a shadow zone, or recirculation zone, for the currents
- suspended sediment may be partly carried across the channel

without deposition, especially for narrow channels, fine sediments, and fast currents, leading to a reduced 'trapping efficiency'

- as the channel fills up with sediment, and becomes shallower, its trapping effect decreases, leading to an exponentially decreasing infill rate
- navigation channels often fill with mud, as well as sand, for which different calculation techniques are required.

The general principles involved in channel infill are discussed by Lean (1980), and sophisticated numerical modelling techniques for more advanced infill calculations are described by Van Rijn (1985) and Miles *et al.* (1995).

The methodologies described in this case study for infill of broad navigation channels can be applied to infill of other broad excavations (e.g. for submerged tube tunnels and cooling water ducts), and for narrow temporary trenches for burial of pipelines, outfalls and cables. The general principles can also be applied in an inverse sense to the dispersion of underwater spoil heaps and sediment storage heaps.

References

Abbott, M. B. and Price, W. A., eds. (1994). *Coastal, Estuarial and Harbour Engineers' Reference Book*, E. & F. N. Spon, London.

Ackers, P. and White, W. R. (1973). Sediment transport: a new approach and analysis. *Proc. ASCE*, **99** (HY11), 2041–60.

Ashida, K. and Michiue, M. (1972). Study on hydraulic resistance and bedload transport rate in alluvial streams. *JSCE*, Tokyo, **206**, 59–69.

Bagnold, R. A. (1963). Mechanics of marine sedimentation, in: *The Sea: Ideas and Observations*, vol. 3, ed. M. N. Hill, pp. 507–28, Wiley-Interscience, New York.

Bailard, J. A. (1981). An energetics total load sediment transport model for a plane sloping beach. *J. Geophys. Res.*, **86**, (C11), 10938–54.

Barltrop, N. D. P. (1990). Revision to the UK DEn guidance notes, in: *Water Wave Kinematics*, eds A. Tørum, and O. T. Gudmestad, pp. 233–46. Kluwer Academic Publishers, Dordrecht.

Battjes, J. A. and Stive, M. J. F. (1985). Calibration and verification of a dissipation model of random breaking waves. *J. Geophys. Res.*, **90**, (C5), 9159–67.

Bettess, R. (1985). Sediment transport under waves and currents. *Report No. SR 22*, HR Wallingford.

BGS (1987). Sea Bed Sediments around the United Kingdom, 1:1,000,000 map (North sheet and South sheet). (Also at 1:250,000 scale for selected areas). British Geological Survey, Natural Environment Research Council.

Bijker, E. W. (1967). Some considerations about scales for coastal models with moveable bed. *Publ. 50*, Delft Hydraulics Lab.

Brampton, A. H. and Motyka, J. M. (1984). Modelling the plan shape of shingle beaches. POLYMODEL 7 Conf., Sunderland Polytechnic, UK, pp. 219–33.

BSI (1967). Methods of testing soils for civil engineering purposes, *British Standard 1377: 1967*. British Standards Institution.

BSI (1986). British Standard specification for test sieves, *British Standard BS410:1986*. British Standards Institution.

CERC (1984). *Shore Protection Manual*, vols I to III. US Army Corps of Engineers, Coastal Engineering Research Centre, US Govt Printing Office.

Chen, S. F., Chan, R. C., Read, S. M. and Bromley, L. A. (1973). Viscosity of sea water solutions. *Desalination*, **13**, 37–51.

Chesher, T. J. and Miles. G. V. (1992). The concept of a single representative wave, in: *Hydraulic and Environmental Modelling: Coastal Waters*, ed. R. A. Falconer, S. N. Chandler-Wilde, and S. Q. Liu, pp. 371–80. Ashgate, Brookfield, VT.

Christoffersen, J. B. and Jonsson, I. G. (1985). Bed friction and dissipation in a combined current and wave motion. *Ocean Eng.*, **12**(5), 387–423.

CIRIA/CUR (1991). *Manual on the Use of Rock in Coastal and Shoreline Engineering*. Construction Industry Research and Information Association, Special Publ. 83.

Colebrook, C. F. and White, C. M. (1937). Experiments with fluid friction in roughened pipes. *Proc. Roy. Soc., Series A*, **161**, 367–81.

Crickmore, M. J. and Teal, C. J. (1981). Recent developments in pump samplers for the measurement of sand transport, in *Erosion and Sediment Transport Measurement*, Proc. Florence Symp., IAHS Publ. 133, pp. 123–30.

Crickmore, M. J., Tazioli, G. S., Appleby, P. G. and Oldfield, F. (1990). The use of nuclear techniques in sediment transport and sedimentation problems, *Technical Documents in Hydrology, IHP–III Project 5.2* International Hydrological Programme, UNESCO, Paris.

Damgaard, J. S. and Soulsby, R. L. (1997). Longshore bed-load transport. *Proc. 25th Int. Conf. Coastal Eng.*, Orlando, **3**, pp. 3614–3627. ASCE.

Davies, A. G., Soulsby, R. L. and King, H. L. (1988). A numerical model of the combined wave and current bottom boundary layer. *J. Geophys. Res.*, **93** (C1), 491–508.

Davies, A. G., Ribberink, J. S., Temperville, A. and Zyserman, J. A. (1997). Comparisons between sediment transport models and observations made in wave and current flows above plane beds. *Coastal Eng.*, **31**, 163–169.

Dawson, G. P., Johns, B. and Soulsby, R. L. (1983). A numerical model of shallow-water flow over topography, in: *Physical Oceanography of Coastal and Shelf Seas*, ed. B. Johns. pp. 267–320. Elsevier, Amsterdam.

De Vriend, H. J. ed. (1993). Coastal morphodynamics: processes and modelling. Special Issue, *Coastal Eng.*, **21**, (1–3).

De Vriend, H. J. (in preparation). *Mathematical Modelling of Coastal Morphodynamics*. World Scientific Publishing, Singapore, Advanced Series on Ocean Engineering.

Deigaard, R. (1998). Comparison between a detailed deterministic sediment transport model and the Bailard formula. *Proc. Coastal Dynamics '97 Conf.*, University of Plymouth (to appear).

Deigaard, R., Bro Mikkelsen, M. and Fredsøe, J. (1991). Measurements of the bed shear stress in a surf zone. *Progress Report 73*, Inst. Hydrodynamic and Hydraulic Eng., pp. 21–30.

Delo, E. A. and Ockenden, M. C. (1992). Estuarine Muds Manual. *Report SR 309*, HR Wallingford.

Den Adel, H. (1987). Re-analysis of permeability measurements using Forchheimer's equation. *Report CO-272550/56* (in Dutch), Delft Geotechnics.

Draper, L. (1991). *Wave Climate Atlas of the British Isles*. Dept of Energy, Offshore Technology Report OTH 89 303, HMSO, London.

Dyer, K. R. (1986). *Coastal and Estuarine Sediment Dynamics*. Wiley & Sons, Chichester, UK.

Einstein, H. A. (1950). The bed-load function for sediment transportation in open channel flows. *Techn. Bulletin 1026*, US Dept of Agriculture.

Engelund, F. (1953). On the laminar and turbulent flow of ground water through homogeneous sand. *Trans. Danish Academy of Technical Sciences*, (3).

Engelund, F. (1966). Hydraulic resistance of alluvial streams. *J. Hydraul. Div., Am. Soc. Civ. Eng.*, **92**, (HY2), 315–26.

Engelund, F. and Fredsøe, J. (1976). A sediment transport model for straight alluvial channels. *Nordic Hydrology*, **7**, 293–306.

Engelund, F. and Hansen, E. (1972). *A Monograph on Sediment Transport in Alluvial Streams*, 3rd edn. Technical Press, Copenhagen.

Ergun, S. (1952). Fluid flow through packed columns. *Chem. Eng. Progress*, **48**(2), 89–94.

Fisher, K. R. (1993). Manual of Sediment Transport in Rivers. *Report SR 359*, HR Wallingford.

Fredsøe, J. (1984). Turbulent boundary layer in wave-current motion. *J. Hydraul. Eng. ASCE*, **110**, 1103–20.

Fredsøe, J. and Deigaard, R. (1992). *Mechanics of Coastal Sediment Transport*. World Scientific Publishing, Advanced Series on Ocean Engineering, vol. 3.

Fredsøe, J., Andersen, O. H. and Silberg, S. (1985). Distribution of suspended sediment in large waves. *J. Waterway, Port, Coastal and Ocean Eng. Div., ASCE*, **111**(6), 1041–59.

Garcia, M. and Parker, G. (1991). Entrainment of bed sediment into suspension. *J. Hydr. Div., Proc. ASCE*, **117**(4), 414–35.

Gibbs, R. J., Matthews, M. D. and Link, D. A. (1971). The relationship between sphere size and settling velocity. *Journal of Sedimentary Petrology*, **41**(1), 7–18.

Graf, W. H. (1984). *Hydraulics of Sediment Transport*. Water Resour. Publ. Littleton, CO, USA.

Grant, W. D. and Madsen, O. S. (1979). Combined wave and current interaction with a rough bottom. *J. Geophys. Res.*, **84**(C4), 1797–1808.

Grant, W. D. and Madsen, O. S. (1982). Movable bed roughness in unsteady oscillatory flow. *J. Geophys. Res.*, **87**(C1), 469–481.

Grass, A. J. (1981). Sediment transport by waves and currents, *Rep. FL29*, SERC London Cent. Mar. Technol., London, UK.

Guy, H. P., Simons, D. B. and Richardson, E. V. (1966). Summary of alluvial data from flume experiments, 1956–1961. *Prof. Paper 462-J*, US Geological Survey.

Hallermeier, R. J. (1981). Terminal settling velocity of commonly occurring sand grains. *Sedimentology*, **28**, 859–65.

Haugen, D. A., ed. (1973). *Workshop on Micrometeorology*. American Meteorological Society, Boston, MA, USA.

Horikawa, K., ed. (1988). *Nearshore Dynamics and Coastal Processes*. University of Tokyo Press.

HRS (1980). Design of seawalls allowing for wave overtopping. *Report EX 924*, Hydraulics Research Station, Wallingford.

Huynh-Thanh, S. and Temperville, A. (1991). A numerical model of the rough turbulent boundary layer in combined wave and current interaction, in *Sand Transport in Rivers, Estuaries and the Sea*, eds R. L. Soulsby and R. Bettess, pp. 93–100. Balkema, Rotterdam.

IAHR/PIANC (1986). List of sea–state parameters, *Supplement to Bulletin 52*. International Association for Hydraulic Research.

Isobe, M. and Horikawa, K. (1982). Study on water particle velocities of shoaling and breaking waves. *Coastal Eng. Jpn.*, **25**, 109–23.

Jinchi, H. (1992). Application of sandwave measurements in calculating bed load discharge, in *Erosion and Sediment Transport Monitoring Programmes in River Basins*, Proc. Oslo Symp. IAHS Publ. 210, pp. 63–70.

Johns, B., Soulsby, R. L. and Chesher, T. J. (1990). The modelling of sandwave evolution resulting from suspended and bed load transport of sediment. *J. Hydr. Res.*, **28**, 355–74.

Johns, B., Soulsby, R. L. and Jiuxing Xing (1993). A comparison of numerical model experiments of free surface flow over topography with flume and field observations. *J. Hydr. Res*, **31**(2), 215–28.

Kamphuis, J. W. (1991). Alongshore sediment transport rate. *J. Waterway, Port, Coastal and Ocean Eng. Div., ASCE*, **117**(6), 624–40.

Kirkgöz, M. S. (1986). Particle velocity prediction at the transformation point of plunging breakers. *Coastal Eng.*, **10**, 139–47.

Koenders, M. A. (1985). Hydraulic criteria for filters, in *Estuary Physics*. Kew.

Komar, P. D. and Miller, M. C. (1974). Sediment threshold under oscillatory water waves. *J. Sediment. Petrol.*, **43**, 1101–1110.

Kraus, N. C. and Horikawa, K. (1990). Nearshore sediment transport, in *The Sea*, vol. 9B: *Ocean Engineering Science*, eds B. Le Mehauté and D. Hanes, pp. 775–814. Wiley, New York.

Latteux, B. (1996). Techniques for long–term morphological simulation under tidal action. *Report 96NV00031*, Electricité de France/Lab. National d'Hydraulique.

Lean, G. H. (1980). *Estimation of Maintenance Dredging for Navigation Channels*. Hydraulics Research Station, Wallingford, UK.

Madsen, O. S. (1991). Mechanics of cohesionless sediment transport in coastal waters, in *Coastal Sediments '91*, eds N. C. Kraus, K. J. Gingerich and D. L. Kriebel, pp. 15–27. ASCE, New York.

Madsen, O. S. and Grant, W. D. (1976). Sediment transport in the coastal environment. *Report 209*, M.I.T. Ralph M. Parsons Lab.

MAFF (1981). *Atlas of the Seas Around the British Isles*. Ministry of Agriculture, Fisheries and Food, Lowestoft. HMSO.

Meyer-Peter, E. and Müller, R. (1948). Formulas for bed-load transport. *Rep. 2nd Meet. Int. Assoc. Hydraul. Struct. Res.*, Stockholm, pp. 39–64.

Miles, G. V. (1981). *Sediment Transport Models for Estuaries*. Hydraulics Research Station, Wallingford, UK.

Miles, G. V., Mead, C. T., Wild, B. R. and Ramsay, D. L. (1995). A study of methods for predicting sandy sedimentation in large dredged trenches or channels in estuaries and recommended practices. *Report SR 410*, HR Wallingford.

Mogridge, G. R., Davies, M. H. and Willis, D. H. (1994). Geometry prediction for wave–generated bedforms. *Coastal Eng.*, **22**, 255–86.

Muir-Wood, A. M. and Fleming, C. A. (1981). *Coastal Hydraulics*. 2nd edn. The MacMillan Press.

Myers, J. J., Holm, C. H. and McAllister, R. F. (1969). *Handbook of Ocean and Underwater Engineering*. McGraw–Hill Book Co., New York.

Myrhaug, D. (1989). A rational approach to wave friction coefficients for rough, smooth and transitional turbulent flow. *Coastal Eng.*, **13**, 11–21.

Myrhaug, D. (1995). Bottom friction beneath random waves. *Coastal Eng.*, **24**, 259–73.

Myrhaug, D. and Slaattelid, O. H. (1990). A rational approach to wave–current friction coefficients for rough, smooth and transitional turbulent flow. *Coastal Eng.*, **14**, 265–93.

Nairn, R. B. and Southgate, H. N. (1993). Deterministic profile modelling of nearshore processes. Part 2. Sediment transport and beach profile development. *Coastal Eng.*, **19**, 57–96.

Nielsen, P. (1979). Some basic concepts of wave sediment transport. *Series Paper 20*, Inst. of Hydrodynamics and Hydraulic Eng., ISVA, Tech. Univ. Denmark.

Nielsen, P. (1992). *Coastal Bottom Boundary Layers and Sediment Transport.* World Scientific Publishing, Singapore, Advanced Series on Ocean Engineering, vol. 4.

Nikuradse, J. (1933). Strömungsgesetze in rauhen Rohren. VDI Forschungsheft 361, Berlin. English translation as: Laws of flow in rough pipes. Natl. Advisory Comm. Aeronautics, Tech. Mem. 1292, Transl. 1950.

Ockenden, M. C. and Soulsby, R. L. (1994). Sediment transport by currents plus irregular waves. *Report SR 376*, HR Wallingford.

Ozasa, H. and Brampton, A. H. (1980). Mathematical modelling of beaches backed by seawalls. *Coastal Eng.*, **4**(1), 47–64.

Parker, G. and Kovacs, A. (1993). Mynorca: a Pascal program for implementing the Kovacs–Parker vectorial bedload transport relation on arbitrarily sloping beds. *Technical Memorandum M-233*, University of Minnesota, St Anthony Falls Hydraulic Laboratory.

Pender, G., Meadows, P. S. and Tait, J. (1994). Biological impact on sediment processes in the coastal zone. *Proc. Instn Civ. Engrs Wat., Marit., & Energy*, **106**, 53–60.

Prandle, D. (1982a). The vertical structure of tidal currents. *Geophys. Astrophys. Fluid Dynamics*, **22**, 29–49.

Prandle, D. (1982b). The vertical structure of tidal currents and other oscillatory flows. *Continental Shelf Res.*, **1**(2), 191–207.

Raudkivi, A. J. (1990). *Loose Boundary Hydraulics, 3rd edn.* Pergamon Press, Oxford.

Ribberink, J. S. and Al–Salem, A. (1991). Near-bed sediment transport and suspended sediment concentrations under waves. *Preprints of Int. Symp. on the Transport of Suspended Sediments and its Mathematical Modelling*, Florence, Italy, 2–5 September, 1991.

Shepard, F. P. (1963). *Submarine Geology*, 2nd edn. Harper & Row, London.

Shields, A. (1936). Anwendung der Ähnlichkeits-Mechanik und der Turbulenzforschung auf die Geschiebebewegung. Preussische Versuchsanstalt für Wasserbau und Schiffbau, vol. 26, Berlin.

Simm, J. D., Brampton, A. H., Beech, N. W. and Brooke, J. S. (1996). *Beach Management Manual, Report 153*. CIRIA, London.

Sleath, J. F. A. (1970). Wave-induced pressures in beds of sand. *J. Hydraul. Div., Proc. ASCE*, **96** (HY2), 367–78.

Sleath, J. F. A. (1978). Measurements of bed load in oscillatory flow. *J. Waterw. Port Coastal Ocean Eng. Div., Proc. ASCE.*, **104** (WW3), 291–307.

Sleath, J. F. A. (1982). The suspension of sand by waves. *J. Hydr. Res.*, **20**, 439–52.

Sleath, J. F. A. (1984). *Sea Bed Mechanics*. Wiley, New York.

Smith, J. D. and McLean, S. R. (1977). Spatially averaged flow over a wavy surface. *J. Geophys. Res.*, **82**(12), 1735–46.

Soulsby, R. L. (1983). The bottom boundary layer of shelf seas, in *Physical Oceanography of Coastal and Shelf Seas*, ed. B. Johns, pp. 189–266. Elsevier, Amsterdam.

Soulsby, R. L. (1987*a*). Calculating bottom orbital velocity beneath waves. *Coastal Eng.*, **11**, 371–80.

Soulsby, R. L. (1987*b*). The relative contributions of waves and tidal currents to marine sediment transport. *Report SR 125*, HR Wallingford, UK.

Soulsby, R. L. (1990). Tidal-current boundary layers, in *The Sea*, vol. 9B, *Ocean Engineering Science*, eds B. LeMehauté and D. M. Hanes, Wiley, New York.

Soulsby, R. L. (1993). Modelling of coastal processes, in *Proc. 1993 MAFF Conf. of River and Coastal Engineers*, Loughborough, UK, pp. 111–118. Ministry of Agriculture, Fisheries and Food, London, UK.

Soulsby, R. L. (1994). Manual of Marine Sands. *Report SR 351*, HR Wallingford.

Soulsby, R. L. (1995*a*). Bed shear-stresses due to combined waves and currents, in *Advances in Coastal Morphodynamics*, ed. M. J. F. Stive, H. J. de Vriend, J. Fredsøe, L. Hamm, R. L. Soulsby, C. Teisson and J. C. Winterwerp, pp. 4–20 to 4–23. Delft Hydraulics, Netherlands.

Soulsby, R. L. (1995*b*). The 'Bailard' sediment transport formula: comparison with data and models. Op. cit., pp. 2–48 to 2–53.

Soulsby, R. L. and Dyer, K. R. (1981). The form of the near-bed velocity profile in a tidal accelerating flow. *J. Geophys. Res.*, **86**, 8067–74.

Soulsby, R. L. and Smallman, J. V. (1986). A direct method of calculating bottom orbital velocity under waves. *Report SR 76*, HR Wallingford.

Soulsby, R. L. and Wainwright, B. L. S. A. (1987). A criterion for the effect of suspended sediment on near-bottom velocity profiles. *J. Hydraulic Res.*, **25**(3), 341–55.

Soulsby, R. L. and Humphery, J. D. (1990). Field observations of wave-current interaction at the sea bed, in *Water Wave Kinematics*, eds A. Tørum and O. T. Gudmestad, pp. 413–428. Kluwer Academic Publishers, Dordrecht.

Soulsby, R. L. and Whitehouse, R. J. S. W. (1997). Threshold of sediment motion in coastal environments. *Proc. Pacific Coasts and Ports '97 Conf.*, Christchurch, **1**, pp. 149–54. University of Canterbury, New Zealand.

Soulsby, R. L., Hamm, L., Klopman, G., Myrhaug, D., Simons, R. R. and Thomas, G. P. (1993). Wave-current interaction within and outside the bottom boundary layer. *Coastal Eng.*, **21**, 41–69.

Southgate, H. N. (1988). Aspects of computational efficiency in models of nearshore hydrodynamics, in *Proc. Int. Conf. Computer Modelling in Ocean Engineering*, Venice, ed. B. A. Schrefler and O. C. Zienkiewicz, A. A. Balkema, Rotterdam.

Southgate, H. N. (1995a). Prediction of wave breaking processes at the coastline, in *Potential Flow of Fluids, Advances in Fluid Mechanics*, vol. 6, ed. M. Rahman, pp. 109–48. Computational Mechanics Publications, Southampton.

Southgate, H. N. (1995b). The effects of wave chronology on medium and long term coastal morphology. *Coastal Eng.*, **26**, 251–70.

Southgate, H. N. and Nairn, R. B. (1993). Deterministic profile modelling of nearshore processes. Part 1. Waves and Currents. *Coastal Eng.*, **19**, 27–56.

Stive, M. J. F., de Vriend, H. J., Fredsøe, J., Hamm, L., Soulsby, R. L., Teisson, C. and Winterwerp, J. C. (1995). *Advances in Coastal Morphodynamics: An Overview of the G8 Coastal Morphodynamics Project 1992–1995.* Delft Hydraulics, Netherlands.

Stride, A. H., ed. (1982). *Offshore Tidal Sands. Processes and Deposits.* Chapman & Hall, London.

Sumer, B. M., Fredsøe, J. and Christiansen, N. (1992). Scour and vertical pile in waves. *J. Waterway, Port, Coastal and Ocean Eng. Div., ASCE*, **118**(1), 15–31.

Swart, D. H. (1974). Offshore sediment transport and equilibrium beach profiles. *Delft Hydraulics Lab.*, Publ. 131.

Raudkivi, A. J. (1988). The roughness height under waves. *J. Hydr. Res.*, **26**(5), 569–584.

Taylor, P. A. and Dyer, K. R. (1977). Theoretical models of flow near the bed and their implication for sediment transport, in *The Sea*, vol. 6, eds E. D. Goldberg, I. N. McCave, J. J. O'Brien and J. H. Steele, pp. 579–602. Wiley-Interscience, New York, NY.

Terzaghi, K. and Peck, R. B. (1967). *Soil Mechanics in Engineering Practice*, 2nd edn. Wiley, New York.

Tucker, M. J. (1991). *Waves in Ocean Engineering: Measurement, Analysis and Interpretation.* Ellis Horwood, New York.

Van Gent, M. R. A. (1993). Stationary and oscillatory flow through coarse porous media. *Report No. 93-9*, Delft University of Technology, Faculty of Civil Engineering.

Van Rijn, L. C. (1984). Sediment transport: part I: bed load transport; part II: suspended load transport; part III: bed forms and alluvial roughness. *J. Hydraul. Div., Proc. ASCE*, **110** (HY10), 1431–56; (HY11), 1613–41; (HY12), 1733–54.

Van Rijn, L. C. (1985). Mathematical model for sedimentation of shipping channels. *Proc. Int. Conf. Numerical and Hydraulic Modelling of Ports and Harbours*, Birmingham, UK, pp. 181–6. BHRA, Cranfield, UK.

Van Rijn, L. C. (1989). Handbook Sediment Transport by Currents and Waves. *Report H461*, Delft Hydraulics.

Van Rijn, L. C. (1993). *Principles of Sediment Transport in Rivers, Estuaries and Coastal Seas*. Aqua Publications, Amsterdam.

Wang, S. and White, W. R. (1993). Alluvial resistance in the transition regime. *J. Hydraul. Eng., Proc. ASCE*, **119** (HY6), 725–741.

Wen, C. Y. and Yu, Y. H. (1966). Mechanics of fluidization. *Fluid Particle Technology, Chemical Engineering Progress Symposium Series*, **62**(62), 100–11.

White, W. R., Paris, E. and Bettess, R. (1980). The frictional characteristics of alluvial streams: a new approach. *Proc. Inst. Civ. Eng.*, Part 2, **69**, 737–50.

Whitehouse, R. J. S. (1993). Combined flow sand transport: Field measurements. *Proc. 23rd Int. Conf. on Coastal Engineering*, Venice, vol. 3, pp. 2542–2555. ASCE.

Whitehouse, R. J. S. (1995). Observations of the boundary layer characteristics and the suspension of sand at a tidal site. *Continental Shelf Research*, **15**(13), 1549–67.

Whitehouse, R. J. S. (1997). Scour at marine structures. A manual for practical applications, *Report SR 417*. HR Wallingford.

Wilson, K. C. (1966). Bed-load transport at high shear-stress. *J. Hydraul. Div., ASCE*, **92** (HY6), 49–59.

Wilson, K. C. (1989a). Mobile-bed friction at high shear stress. *J. Hydraulic Eng., ASCE*, **115**(6), 825–30.

Wilson, K. C. (1989b). Friction of wave induced sheet flow. *Coastal Eng.*, **12**, 371–79.

Yalin, M. S. (1963). An expression for bed-load transportation. *Proc. ASCE*, **89** (HY3).

Yalin, M. S. (1964). Geometrical properties of sand waves. *J. Hydraul. Div., Proc. ASCE*, **90** (HY5), 105–19.

Yalin, M. S. (1977). *Mechanics of Sediment Transport*, 2nd edn. Pergamon Press, Oxford.

Zanke, U. (1977). Berechnung der Sinkgeschwindigkeiten von Sedimenten, Mitt. des Franzius–Instituts für Wasserbau, **46**(243). Technical University, Hanover.

Zyserman, J. A. and Fredsøe, J. (1994). Data analysis of bed concentration of sediment. *J. Hydraul. Eng., ASCE*, **120**(9), 1021–42.

Index

Critical bed shear-stress, *see* Threshold of motion
Current ripples, 8, 19, 80, 113, 114–120, 123
Current speed, depth-averaged, 46–58, 76, 88, 93

Damgaard and Soulsby, 196–199
Darcy's Law, 37, 38
Darcy–Weisbach resistance coefficent, 53
Davies, 150, 187
Deigaard, 82, 188
Density,
  of sediment, 10, 15, 21, 158, 225
  of water, 9, 10, 14, 21, 22, 25–27, 37–41
Deposition, 6, 17, 19, 203, 230, 233
Depth-averaged current speed, *see* Current speed, depth-averaged
Design wave and tide approach, 18, 214–215
Diffusion, 6, 12, 92, 137–139, 148, 179, 184–190
Diffusivity, 138, 140, 145, 149, 185–187
Dimensionless grain size, 15, 104, 133, 162
Dispersion relation, 70
Drag coefficient, 53, 55, 59, 93, 143, 159, 175, 183, 192, 231
Dredged channels, 5, 173, 190, 204, 230–233
Dunes, 16, 35, 53, 57, 81, 113–119, 123–126, 174, 175
Duration, 65

Eddy diffusivity, 130, 140, 145, 149, 185–187
Einstein, 58, 160
Engelund, 39, 40, 42, 125, 126, 175, 177
Engelund and Hansen, 167, 174, 177, 179
Entrainment, 6, 8, 40, 58, 92, 123, 138, 179
Equivalent monochromatic wave, 17, 70, 80
Erosion, 4, 6, 8, 17, 19, 203, 207

Errors, 5, 12, 20–22, 78, 195, 199

Fall velocity, *see* Settling velocity
Fetch, 65
Field measurements, 48, 50, 59, 83, 182, 190
Fluidisation, 37–42, 136
Folk classification, 30, 31, 33
Forchheimer's equation, 38, 40
Form drag, 11, 58, 59, 81, 123, 138, 147
Fredsøe, 90, 95, 186
Fredsøe and Deigaard, 3, 11, 70, 79, 132
Friction factor, *see* Wave friction factor
Friction velocity, 9, 16–19, 46, 53, 58, 59, 138, 175

Geometric standard deviation, 28
Grain-related bed shear-stress, 11, 54, 58, 59, 81
Grain-size distribution, 14, 28–30, 33, 40, 48
Grant and Madsen, 82, 90, 93, 95, 121, 123, 124, 127
Grass, 182, 184
Gravel, 3, 6, 13, 14, 28, 31, 33, 39, 47, 48, 105, 107, 115, 189

Half-cycle transport, 163–166
Hindered settling, 134, 136
H_s-T_z scatter diagram, 17, 68, 216
Hydrodynamically rough, 47, 48, 93
Hydrodynamically smooth, 47

Incipient motion, *see* Threshold of motion
Initiation of motion, *see* Threshold of motion

JONSWAP spectrum, 66, 68, 70, 73, 75, 76, 191, 231

Kinematic viscosity, 14, 21, 25–27, 37, 41, 47, 133

Littoral drift, 193
Locally-generated sea, 65